ENVIRONMENTAL SOCIOLOGY

A social constructionist perspective

John A. Hannigan

London and New York

First published 1995
by Routledge
11 New Fetter Lane, London EC4P 4EE

Simultaneously published in the USA and Canada
by Routledge
29 West 35th Street, New York, NY 10001

Typeset in Bembo by Michael Mepham, Frome, Somerset
Printed and bound in Great Britain by
T.J. Press (Padstow) Ltd, Padstow, Cornwall

British Library Cataloguing in Publication Data
A catalogue record for this book is available from the
British Library

Library of Congress Cataloguing in Publication Data
A catalogue record for this book has been requested

ISBN 0–415–11254–0 (hbk)
ISBN 0–415–11255–9 (pbk)

To Ruth

CONTENTS

PREFACE AND
ACKNOWLEDGEMENTS

As is the case with sociology in general, I came to the area of environmental studies later rather than sooner. Indeed, there were several missed opportunities of note earlier in my academic career. In 1970, my first year of graduate work, a good friend and fellow sociology student at the University of Western Ontario, William Trotter, became active in the newly founded environmental group, Pollution Probe, later going on to forge a professional career in the environmental field. I wish now that I had looked more closely at what Bill was up to in those days. In 1973, I went on to doctoral studies at the Ohio State University where I was a research associate at the Disaster Research Center. Despite an obvious synergy between disaster research and environmental sociology, other specialties, notably, collective behaviour and the sociology of complex organisations, largely shaped the Center's perspective during this period. I wish I had made the connection.

As it was, I did not begin to take an active interest in environmental issues and outlooks until the late 1980s. In the first instance, my wife and I became active for several years in a variety of environmental conflicts in our community, first as part of a grassroots campaign to shut down an outdated garbage incinerator and then to oppose the construction of sewage detention tanks along the beachfront (see Chapter 5). Fortuitously, in 1989, Raimondo Strassaldo, who had read an article of mine on what subsequently became known as the 'New Social Movements', invited me to participate in a conference in Udine, Italy, on 'Environmental Constraints and Opportunities in the Social Organization of Space'. At this conference I had the chance to meet and talk with Riley Dunlap, Michael Redclift and other major contributors to the emerging field of environmental sociology. From that point on I was hooked. Thank you, Raimondo, for providing this marvellous opportunity. I have since greatly profited from attending other such conferences, notably the 1992 Symposium, 'Current Developments in Environmental Sociology' at the Woud-schoten conference centre in the Netherlands.

In the present, a sundry cast has helped to shape this book and move it along to completion. I have sharpened my thinking about the environment–society relationship from conversations with colleagues David Tabara (Barcelona), Dave Tindall (Vancouver) and Shelly Ungar (Toronto). Similarly, I have had a useful opportunity to work out some of my ideas in my graduate level course, 'Environmental Sociology', with a small but keen group of students: Slobodan Drakulic, Ivanca Knezevic, Joel Lau, Guadaloupe Mateos Marcos and Mary Beth Raddon. Chris Rojek, former senior sociology editor at Routledge, scouted my proposal on a trip to Canada and brought it in short order to the editorial committee. Ann Gee has been a helpful and efficient contact at Routledge, making sure that this project never got bogged down and slipped into next year's publication list. Angie Doran, managing production editor, efficiently saw the book through to publication. Steven Yearley, editor of Routledge's 'Environment and Society' series, made many valuable suggestions for revising and improving the initial draft.

In Toronto, my wife, Ruth, was instrumental in urging me to press ahead with this project and to submit it for international publication. Furthermore, she adroitly held things together at home on the many week nights and Saturdays when I was entombed in the university library, despite an especially hectic year developing her own career. Both our sons assisted in turning out the book: TJ in the computer set-up and printing of the manuscript and Tim in typing a portion of the bibliography. Our girls, Maeve and Olivia, helped too by always being full of enthusiasm and good cheer. At Scarborough College, Carole Tuck and Audrey Glasbergen guided me through some last-minute crises with photocopying and printing. Thank you one and all.

ABBREVIATIONS

AEC	Atomic Energy Commission (US)
Alpac	Alberta–Pacific Forest Industries
AOU	American Ornithologists Union
BEN	Black Environmental Network (UK)
bST	Recombinant Bovine Somatotropin
BUND	*Bund für Unwelt und Naturschutz Deutschland* (Germany)
CASAC	Clean Air Scientific Advisory Committee (US)
CAST	Council of Agricultural Science and Technology
CCAR	Canadian Coalition on Acid Rain
CEGB	Central Electricity Generating Board (UK)
CFCs	Chloroflurocarbons
CIPRs	Collective Intellectual Property Rights
CITES	Convention on International Trade in Endangered Species of Wild Fauna and Flora
D.o.E	Department of the Environment (UK)
EACN	European Air Chemistry Network
EC	European Community
EPA	Environmental Protection Agency (US)
EU	European Union
FDA	Food and Drug Administration (US)
FET	Foundation on Economic Trends (US)
FGD	Flue Gas Desulphurisation
GMAG	Genetic Manipulation Advisory Group (UK)
HEP	Human Exemptionalism Paradigm
IDFA	International Dairy Foods Association
INA	Ice Nucleated Bacterium
IUCN	International Union for Conservation of Nature
NAACP	National Association for the Advancement of Colored People

NAS	National Academy of Sciences (US)
NASA	National Aeronautics and Space Administration (US)
NCAC	National Clean Air Coalition (US)
NEP	New Ecological Paradigm
NGO	Non-Governmental Organisation
NIH	National Institutes of Health (US)
NOAA	National Oceanic and Atmospheric Administration (US)
NSF	National Science Foundation (US)
NSMs	New Social Movements
NUM	National Union of Mineworkers (UK)
ONAC	Office of Noise Abatement and Control (US)
OPEC	Organisation of Petroleum Exporting Countries
PVPA	Plant Variety Protection Act (US)
SCB	Society for Conservation Biology
SST	Supersonic Transport Airplane
UCC	United Church of Christ
UFW	United Farm Workers
UNCED	United Nations Conference on Environment and Development
UNEP	United Nations Environmental Programme
UNESCO	United Nations Economic, Social and Cultural Organization
USAID	US Agency for International Development
USDA	US Department of Agriculture
WWF	World Wildlife Fund

INTRODUCTION

When I was young, I lived in Windsor, Ontario, a lunchbucket Canadian community adjacent to the much larger American metropolis of Detroit. Nicknamed the 'Motor City', Detroit was one of the old industrial cities of the north-eastern United States, the site of numerous automotive, steel, cement and other manufacturing plants as well as a prominent coal-fired electrical power plant. Windsor also had an industrial base with auto plants (Chrysler, Ford), a distillery and a salt-mine. The two cities were separated by the Detroit River, in the 1950s a busy transportation corridor for lake freighters loaded down with iron ore, grain and other cargoes.

Along the shoreline, a block from our house, was a vacant lot with a permanent 'NO SWIMMING' sign driven into the ground, a stark reminder of the polluted quality of the water. Not only was the river befouled by continual industrial, chemical and shipping-related discharges but, as Congressman (later President) Gerald Ford once pointed out about Detroit, the city that was parent to the automobile was served by a 'horse and buggy sewage plant' which only gave primary treatment to a massive waste stream before discharging it into local waters (Ashworth 1986: 141). The air too was frequently clouded by emissions from the power plant and the factories, especially those located on the aptly named Zug Island. One summer, just after we had moved to another city, it even snowed in the south-western section of Windsor; the snow turned out to be an airborne discharge from one of the industrial sources on the American side of the river.

While most Windsorites were fully aware of the less than pristine state of the air and water, no campaigns were introduced demanding a major clean-up. People accepted the pollution (it was not called that then) as an unpleasant fact of life about which they could do little. If anything, Windsor residents worried more about the possibility of a downturn in the booming auto industry which would revive the job lay-offs that they had experienced from time to time in the past.

Ecology was an unknown word in the school classrooms. In fact, the one primary school project which I still recall involved clipping coupons from a special promotional supplement published by a series of corporate sponsors, sending them in to receive further information, and finally writing a composition on that particular consumer product (I chose the soft drink Orange Crush).

Compare this with the experience of Charles Rubin, an American political scientist, who grew up in Cleveland, Ohio, six years later. In his book, *The Green Crusade*, Rubin recounts how he was a member of the first generation educated to 'look around and see "environmental problems" as such' (1994: 8). Like Detroit, Cleveland is an industrial city along the Great Lakes which has long been deluged with pollution: it is celebrated (or perhaps notorious) for being the site of the Cuyahoga River which became so befouled with industrial waste that it once caught fire. However, by the time Rubin reached the fifth grade in 1964–5, the environmental movement had already been born. His teacher taught her pupils about plant succession, the eutrophication of lakes, weather and birds. Their class project was to 'save' Lake Erie. Rubin even remembers attending a birthday party, the highlight of which was a visit to the local sewage treatment plant.

What this illustrates is that public concern about the environment is by no means automatic even when conditions are visibly bad. In 1958–9, Windsor residents did not define the existence of noxious and potentially harmful pollutants in the air and water as especially problematic or actionable despite what even today would be regarded as unacceptable levels. By contrast, in 1964–5, at least a minority of Cleveland residents had done so and had begun to communicate this concern; so too eventually did Windsorites. Thirty years later, the Ontario provincial government even took the unprecedented step of going to court in the state of Michigan to halt the construction of a jumbo waste incinerator in Detroit, which, it was feared, would seriously mar the air quality of the border cities area.

Clearly, then, environmental concern is not constant but fluctuates over time, rising and falling in prominence. Furthermore, environmental problems do not materialise by themselves; rather, they must be 'constructed' by individuals or organisations who define pollution or some other objective condition as worrisome and seek to do something about it. In this regard, environmental problems are not very different from other social problems such as child abuse, homelessness, juvenile crime or Aids. From a sociological point of view, the chief task here is to understand why certain

2

conditions come to be perceived as problematic and how those who register this 'claim' command political attention in their quest to do something positive.

In this book, I adopt an explicitly social constructionist perspective in order to examine the rise, and in some cases fall, of a wide range of environmental problems from wildlife decimation in the nineteenth century to ozone depletion in the late twentieth century. Along the way, I focus on the role of two major societal institutions – science and the mass media – in constructing environmental risks, knowledge, crises and solutions. Three contemporary environmental problems are analysed in depth: acid rain, biodiversity loss and biotechnology (in the form of the genetically altered hormone, bST) as an environmental risk.

Unlike a spate of recent books which attempt to debunk environmentalism and its system of beliefs (Bailey 1993; Fumento 1993; Lewis 1992; Rubin 1994), the object of this analysis is not to discredit environmental claims but rather to understand how they are created, legitimated and contested. Like Mazur and Lee (1993: 714), I am not by any means attracted to an extreme constructionist position which insists that the global ensemble of problems is purely a creation of the media (or science or ecological activists) with little basis in objective conditions. On the contrary, perhaps as a result of having witnessed hardcore urban pollution early in life, I fully recognise the mess which we have created in the atmosphere, the soil and the waterways.

At the same time, in the course of peering through a constructionist lens at various environmental issues and problems, some things which I took for granted before were cast in a new perspective while other significant 'facts' came to light. Rachel Carson, although by no means the first scientist to proclaim the dangers of pesticides for the food chain, was the most successful. Acid rain was first identified as far back as the nineteenth century but not acted upon until the 1960s, when an obscure but enterprising Swedish biologist serendipitously linked it to the death of fish in the lakes of Scandinavia. There really is no ozone hole as such but rather a thinning in concentration; the image of the hole was scientifically constructed to make the situation more dramatic and understandable. 'Endangered' species of plants and animals are not always universally threatened but only in a particular regional habitat. The Gran Chaco, a mostly arid lowlands plain which constitutes Latin America's second largest ecosystem after the Amazon, is disappearing at a much faster rate than the tropical rainforest but, unlike the latter, remains virtually unknown internationally (Angier 1994:13–19).

As we shall see, the constructionist perspective is by no means the exclusive preserve of sociologists. Political scientists, for example, have been at the forefront of this approach from the fine work on issue building and agenda setting in the domestic political arena carried out by Kingdon (1984), Solesbury (1976) and Walker (1977, 1981) to more recent research on the role of 'epistemic communities' (Haas 1990, 1992) in the legitimation and mobilisation for action on transborder environmental problems. Similarly, environmental historians have unearthed a wealth of data on the social construction of nature both during the earlier conservation era (Nash 1967; Schrepfer 1983; Worster 1977) and, more recently, in the rise of the modern environmental movement (Gottlieb 1993).

Nevertheless, I believe that sociologists can play an important role in developing the constructionist project in two ways. First, by drawing on a critical mass of theory from Berger and Luckmann, Blumer, Gamson, Gusfield, Kitsuse and others, sociological analysts are well placed to conceptualise the process of environmental claims-making within a wider framework of reality construction, frame analysis and symbolic interactionism. Second, building on the Weberian tradition, sociologists should weld the concept of 'power' to that of social construction to chart how and why some claims are accorded legitimacy and others are rebuffed. The recent outpouring of sociological attention to the linked phenomena of 'environmental racism' and 'environmental justice' (Bryant and Mohai 1992; Bullard 1990; Capek 1993; Hofrichter 1993) is one healthy example of how it is possible for this approach to yield important insights.

Let us start our discussion of social constructionism and the environment by briefly examining why sociology failed to centrally address environmental issues much earlier in its developing career as a social science discipline.

1

ENVIRONMENTAL SOCIOLOGY

Issues and theoretical approaches

'Earth Day 1970' is often said to represent the debut of the modern environmental movement. Starting as a modest proposal for a national teach-in on the environment, it grew into a multi-faceted event with millions of participants. What most distinguished Earth Day, however, was its symbolic claim to be 'Day one' of the new environmentalism, an interpretation which was widely embraced by the American mass media which afforded the environmental issue instant and widespread recognition (Gottlieb 1993:199).

When Earth Day inaugurated the 'Environmental Decade' of the 1970s, sociologists found themselves without any prior body of theory or research to guide them towards a distinctive understanding of the relationship between society and the environment. While each of the three major classical sociological pioneers – Emile Durkheim, Karl Marx and Max Weber – arguably had an implicit environmental dimension to their work, this had never been brought to the fore, largely because their American translators and interpreters favoured social structural explanations over physical or environmental ones (Buttel 1986: 338). From time to time, isolated works pertaining to natural resources and the environment had appeared, mostly within the area of rural sociology, but these had never coalesced into a cumulative body of work. In a similar fashion, social movement theorists gave short shrift to conservation groups, leaving historians to explore their roots and significance.

To comprehend why this situation arose, it is necessary to consider how both geographical and biological theories of social development and social change lost their predominance when sociology emerged as a distinctive discipline in the early twentieth century.

THE FAILURE OF GEOGRAPHICAL AND
BIOLOGICAL DETERMINISM

In the nineteenth century, the effects of the geographical environment on the human condition was a topic of considerable scholarly interest. Perhaps the leading geographical determinist was the British historian, Henry Thomas Buckle, author of *The History of Civilization in England*. Buckle was greatly influenced by the writing of the seventeenth-century French philosopher, Montesquieu, and by several German geographers, notably Karl Ritter. His central thesis was that human society is a product of natural forces, and is therefore susceptible to a natural explanation (Bierstedt 1981: 2). Buckle believed that the influence of the geographical environment is most direct and therefore strongest upon 'primitive' people but declines with the advance of modern culture. He ascribed particular sociological significance to the visual aspect of nature: if the natural environment is awe-inspiring in its beauty or terrifying in its power of destruction, it overdevelops the imagination; if it is less formidable, a more rational intelligence prevails. England, with its gently rolling hills and domesticated farm animals, represented a prime example of the latter.

Buckle's geographical theory of social change was widely read and quite influential in intellectual circles in the nineteenth century (Timasheff and Theodorson 1976: 93). For example, the economist, Thomas Nixon Carver used *The History of Civilization in England* in his sociology course at Harvard long before that university had a formal department of sociology, while William Graham Sumner, widely regarded as the first American sociologist, became interested in Buckle's work while studying theology at Oxford (Bierstedt 1981: 2).

A second leading geographical determinist was Ellsworth Huntington. In his principal sociological works, *Civilization and Climate*, *World Power and Evolution* and *The Character of Races*, Huntington attempted to establish a series of correlations between climate and health, energy, and mental processes such as intelligence, genius and willpower. Having divined the parameters of an 'optimal climate' he then attempted to prove that the rise and fall of entire civilisations such as that of ancient Rome follows the shift of the climatic zones in historical periods.[1]

In assessing the worth of this 'geographical school', Sorokin (1964 [1928]: 192–3) refers to its fallacious theories, its fictitious correlations and its overestimation of the role of the geographical environment, but at the same time he cautions that 'any analysis of social phenomena which does not take into consideration geographical factors is incomplete'.

The natural world also entered into early sociological discourse through the Darwinian concepts of 'evolution', 'natural selection' and the 'survival of the fittest'. In Darwin's theory, those plants and animals which are best suited to adapt to their environment survive, while those which are less well equipped perish. The survivors pass on their advantages genetically to subsequent generations. Darwinism was seized upon by many of the early conservative sociological thinkers who applied its principles (not always accurately) to the human context (see Hofstadter 1959). The most prominent social Darwinist was the English social philosopher, Herbert Spencer, who proposed an evolutionary doctrine which extended the principle of natural selection to the human realm. Spencer bitterly opposed any suggestion that society could be transformed through educational or social reform; rather, he believed that, if left alone, progress would evolve in a gradual fashion.

Sumner was Spencer's greatest academic disciple in America, introducing his own concept of the 'competition of life' whereby humans struggle not just with other species for survival in the natural universe but also with each other in a social universe. Applying his theory to the *laisser-faire* capitalism of the day, Sumner legitimated the triumph of the 'robber barons', millionaire industrialists who made their money in banking, railroads and utilities through sharp and ruthless dealing. They were, Sumner claimed, 'a product of natural selection' who would move society forward on the road to progress.

Both these 'single factor theories of social change' (Bierstedt 1981: 487) were rejected by mainstream sociology for largely the same reasons. By the 1920s the evolutionary *laisser-faire* doctrines of the nineteenth century had given way to a new emphasis on social planning and social reform. 'Meliorism' – the deliberate attempt to improve the well-being of members of society – flew in the face of these social theories which viewed social causation as unalterable, whether due to geography or biology.

Furthermore, by this time the foundation of sociological theory had shifted. Many sociologists had come to accept psychology as the foundation of sociology in place of physics or biology (Timasheff and Theodorson 1976: 188). This was especially evident in the social psychological tradition established by Mead, Cooley, Thomas and other American 'symbolic interactionists' who emphasised that the reality of a situation lies entirely in the definition attached to it by participating social actors. This definition, in turn, was socially shaped, as in Cooley's concept of the 'looking glass self'. Physical (and environmental) properties became relevant only if they were perceived and defined as relevant by the actors (Dunlap and Catton 1992/3: 267).

Increasingly, the failure of social Darwinism, and to a lesser extent the inability of geographic determinism to ever get off the ground, led to a strong aversion to explanations which used biological–environmental explanations (Buttel and Humphrey, forthcoming). This opposition to biological currents was similarly evident in sociology's sibling discipline, anthropology.

After his move to the United States in 1989, Franz Boas, widely recognised as the founder of American cultural anthropology, responded to the rising tide of eugenics, 'scientific' racism and other manifestations of biological determinism by elevating culture to a primary role in individual and societal development, dwarfing both the physical environment and biological inheritance. This emphasis on cultural processes was carried on in this century by such well-known anthropologists as Margaret Mead and Ruth Benedict (Benton 1991: 13). Culture, in fact, came to be valued as the key influence on all aspects of human society.

Ironically, while sociology rid itself of biological explanation, it hung on to a distinctly biological terminology. Functionalism, the leading sociological theory of the 1950s in America, carried forward Durkheim's idea that society constituted a social 'organism' which was constantly having to adapt to the outside social and physical environment. Its equilibrium or steady state could be knocked out of kilter by various disruptive events but, ultimately, it would return to normal just as the human body recovers from a fever. Dickens (1992) has noted that functionalist theorists, especially their dean, Talcott Parsons, might have gone further and actually developed a theory of social evolution in an environmental context which stressed how biological inheritance permitted humans to both adapt to the natural world and to change it. This potential, however, was never developed, leaving environmental factors as marginal elements in sociological explanation.

SOCIOLOGISTS AS 'HUCKSTERS' FOR DEVELOPMENT AND PROGRESS

A second explanation for sociological foot-dragging on environmental matters pertains to the world-view of sociologists themselves.

In a steady stream of papers and articles from the late 1970s on, American rural sociologists William Catton and Riley Dunlap argued that the vast majority of sociologists share a fundamental image of human societies as exempt from the ecological principles and constraints which govern other species. While sociologists are inclined to favour the use of social engineering to achieve such goals as equality, they nevertheless fully accept the possibility of endless growth and progress via continued scientific and

8

technological development while ignoring the potential constraints of environmental phenomena such as climate change (Dunlap and Catton 1992/3: 270).

Some sociological specialities went even further, actively becoming advocates, and even 'hucksters', for the benefit of technological innovation and economic development. Nowhere was this more evident than in the sociological literature on modernisation which was influential for two decades between 1955 and 1975.

Two works in particular stand out in the study of the modernising process: Inkeles and Smith's *Becoming Modern* (1974) and Lerner's *The Passing of Traditional Society* (1958).

For Inkeles and Smith, modernisation denotes both a societal and personal transformation. At the societal level, modernisation is conceptualised as a process of nation and institution building. In the 1960s, the 'decade of development', many Third World nations failed to make their entry into the modern world, sliding backwards into tribalism and ethnic conflict. Newly liberated from colonialism, these emerging countries were said to be 'hollow shells, lacking the institutional structures which make a nation a viable and effective socio-political and economic enterprise' (Inkeles and Smith 1974: 3).

Inkeles and Smith argue that the primary reason for this failure to modernise was that individual members of the community were psychologically trapped in the past, unable to transcend traditional ways of thinking to become modern personalities. Modern citizens possessed a panoply of skills: they could keep to fixed schedules, observe abstract rules, adopt multiple roles and empathise with others. They were optimistic, opinionated, open to new experience and consumers of information. These qualities are not inborn but must be acquired through life experience.

While some of this socialisation in modern ways could be carried out by the educational system, it is the factory, Inkeles and Smith conclude, that is the true 'school in modernity'. The factory, they observed, is the epitome of the institutional pattern of modern civilisation. It functions as a powerful model for rural migrants from traditional settings inculcating, among other qualities, a sense of efficacy, a readiness for innovation and an openness to systematic change, respect for subordinates and the importance of planning and time.

To Daniel Lerner (1958), the key correlate of developing modernity was the media's role in establishing a psychological openness to change among peasant populations. In particular, the media were depicted as fostering a sense of 'empathy' – the ability to imagine change by putting oneself in the shoes of those in society who were engaged in playing roles (e.g. social leader) other than one's own.

In the ascent to modernity the influence of the physical environment was downgraded. Inkeles and Smith (1974:22) observed that a key part of developing a sense of modern efficacy lay in the ability to develop a potential 'mastery' of nature. In their questionnaire, administered to a thousand men in six developing countries, they pose this question:

Which of the following statements do you agree with more?

1 Some people say that man will some day fully understand what causes such things as floods, droughts and epidemics.
2 Others say that such things can never fully be understood by man.

The respondent who was more committed to advancing his own goals rather than being dominated by natural forces would respond positively to the first statement. This view of nature is of course the antithesis of the ecological ethic which stresses that human beings have no inherent claim to domination over nature but must simply coexist with other species on the earth.

One of the few commentators on modernisation in the 1960s to recognise the potential constraints imposed by the environment was Clifford Wharton, an agricultural economist, who noted the special characteristics of agriculture which related to climate, soil and other inputs. 'Bananas do not grow in Alaska (except perhaps in a hothouse)' but 'a shoe factory in Tokyo need not be different from one in São Paulo', Wharton (1966) observed, concluding that agriculture was far more subject to environmental factors than other forms of economic development.

Mesmerised by the benefits of economic development and its sidekick, individual modernity, most sociologists, by contrast, either completely ignored the natural environment or viewed it as something to be overcome with grit and ingenuity.

That is not to say that there were not isolated critics of the pro-development paradigm, especially within the ranks of Marxist sociology. But, like religion, they tended to see the environment as a distraction from the necessity of class struggle. Even where the seriousness of environmental destruction was acknowledged, left-wing critics were inclined to focus on the class and power relations underlying this crisis rather than on factors relating more directly to the environment itself (see Enzenberger 1979). Insomuch as Marxism eventually came to dominate social theory in some important regions of post-war European social theory, this resulted in the further exclusion of environmental issues from the discipline of sociology (Cotgrove 1991; Martell 1994).

TOWARDS AN ENVIRONMENTAL
SOCIOLOGY: 1970 TO 1995

In the past quarter-century, sociologists have shown a more conce~ [1] interest in the environment than was the case in the past. By the mid-19 all three of the national sociological associations in the United States (American Sociological Association, Rural Sociological Association, Society for the Study of Social Problems) had established sections relating to environmental sociology (Dunlap and Catton 1979). Special issues on environmental topics have appeared in a number of sociological journals, for example, *Sociological Inquiry* (1983), *Journal of Social Issues* (1992), *Qualitative Sociology* (1993), *Social Problems* (1993), *Canadian Review of Sociology and Anthropology* (1994). *The Annual Review of Sociology* has twice (1979 and 1987) featured essays on environmental sociology as well as pieces on energy and on the sociology of risk. Furthermore, in 1993, Shirley Laska's presidential address to the Southern Sociological Society was entitled 'Environmental sociology and the state of the discipline'(Laska 1993).

In Europe, stimulated by the emergence of the 'Greens' as a political force, much of the early work on environmental topics dealt with environmentalism and the environmental movement (Dunlap and Catton 1992/3: 273). One exception to this was in the Netherlands where nodes of activity in environmental sociology formed early on around questions pertaining to agriculture and risk assessment. In Britain, past interest in the environment has been explicitly theoretical, weighing the relationship between society and nature against classical sociological perspectives on social class and industrialism. More recently, empirical research on environmental topics has begun to flourish in the UK, in part due to the stimulus provided by the Global Environmental Change programme set up by the Economic and Social Research Council (ESRC), which has underwritten an impressive array of conferences, study groups and symposia.

Interest has also begun to build internationally. In 1992, the Environment and Society group within the International Sociological Association merged with a second Social Ecology group to form Research Committee 24 Environment and Society, with a combined membership of over two hundred members, many of whom are environmental sociologists. At the 1994 World Congress of Sociology in Bielefeld, Germany, seventeen sessions were scheduled encompassing a total of 114 papers on matters relating to the environment and society, while at the 1993 Centennial Congress of the International Institute of Sociology in Paris there were several working sessions on the topic of 'Environmental risks and disasters'. How should this considerable evidence of scholarly activity be assessed?

In the late 1970s, Catton and Dunlap undertook a crusade to convert sociologists to their New Ecological Paradigm (NEP)[2] which was meant to cross-cut the established divisions within sociological theory. This new paradigm was an academic analogue of green thinking in general, advocating an approach which was less 'anthropocentric' (human-oriented) and more 'ecocentric' (humans are only one of many species inhabiting the earth). Buttel (1987: 466) describes their efforts as nurturing a set of 'lofty intentions' wherein environmental sociologists 'sought nothing less than the re-orientation of sociology toward a more holistic perspective that would conceptualise social processes within the context of the biosphere'. Catton and Dunlap now acknowledge that they failed in this endeavour but claim that they never fully expected to achieve this kind of disciplinary conversion (1992/3: 272). Both Buttel and Catton and Dunlap have observed that the environmental sociology field faltered during the Reagan era. However, while Buttel pessimistically refers to environmental sociology as having become just 'another sociological specialization', Catton and Dunlap suggest that the resurgence of interest in environmental issues in the 1990s, especially those which are global in scope, has stimulated renewed interest in environmental sociology in the United States as well as internationally.

What the field still lacks is a seminal work which could lift environmental sociology into the mainstream of debate in the broader field of sociology. One such theoretical soliloquy is Ulrich Beck's book, *The Risk Society* (see Chapter 10 for a detailed assessment). Beck, a sociologist of institutions, has approached the subject of environmental risks more from the perspective of a macro-sociology of social change (Lash and Wynne 1992: 8) than from a paradigm that is rooted specifically in environmental sociology. Nevertheless, Beck's argument has been widely noticed and has provoked considerable discussion both within and beyond the confines of environmental sociology. By contrast, Catton and Dunlap's HEP (Human Exemptionalism Paradigm)–NEP (New Ecological Paradigm) distinction – the most influential theoretical insight within the area of environmental sociology to date – has failed to generate much excitement outside of this new specialty area and its siblings in psychology, political science and environmental education.

In the meantime, it probably makes sense to embrace Elizabeth Shove's (1994) notion that sociologists can make a positive contribution to the environmental debate by both *incorporating* and *engaging*. The former suggests that pockets or niches of environmental research can enrich

12

mainstream sociological theory even if they do not as yet have the capacity to transform the discipline as a whole. The latter recognises that there is much to gain in applying the sociological imagination to the extra-disciplinary study of contemporary environmental issues; for example, through political economy models or via the sociology of science and knowledge. Unfortunately, sociologists far too often end up as 'underlabourers' in this endeavour, being viewed as supporting actors in a cast dominated by natural scientists and environmental policy-makers.

THEORETICAL APPROACHES TO
ENVIRONMENTAL SOCIOLOGY

A continuing problem for sociologists researching the environment has been to define what constitutes the main object of study. In their 1979 review article, Catton and Dunlap pinpointed the distinctive core of the field as a 'new human ecology' which focuses on the interaction between the physical environment and social organisation and behaviour. When it came to identifying areas of research in environmental sociology, however, they allowed a number of topics (the 'built' environment, natural disasters, social impact assessment) which seemed to stretch the parameters of the field rather than to narrow them. Eight years later, Buttel (1987) cited five key areas of environmental sociological scholarship: (1) Catton and Dunlap's 'new human ecology'; (2) environmental attitudes, values and behaviours; (3) the environmental movement; (4) technological risk and risk assessment; and (5) the political economy of the environment and environmental politics, but they did not attempt to locate any common theoretical thread in this diverse menu. A major consequence of this blend of topics and approaches is the absence of any solid consensus on a theoretical base for environmental sociology; the ambiguity resulting from a theoretical vacuum has significantly undermined the legitimacy of this specialty area (Cable and Cable 1995: vii).

In my estimation, there are two distinct problems that are centrally addressed in the existing literature on environmental sociology: (1) the causes of environmental destruction, and (2) the rise of environmental consciousness and movements. Rather than classifying theories of 'environmentalism' or those describing the 'society–nature relationship' (Martell 1994), it makes more sense to discuss past theoretical approaches to the environment and society separately under each of these two headings.

Causes of environmental destruction

In explaining the causes of widespread environmental destruction on our planet, two primary approaches have been offered: the ecological explanation as embodied in Catton and Dunlap's model of 'competing environmental functions', and the political economy explanation as found in Alan Schnaiberg's concepts of the 'societal–environmental dialectic' and the 'treadmill of production'. As Buttel (1987: 471) has noted, both approaches view social structure and social change as being reciprocally related to the biophysical environment but the nature of this relationship is depicted very differently.

Ecological explanation

The ecological explanation for environmental destruction has its roots in the field of 'human ecology' which was dominant within urban sociology from the 1920s to the 1960s.

Urban ecology was first pioneered by Robert Park and his colleagues at the University of Chicago in the 1920s. Park was well acquainted with the work of Darwin and his fellow naturalists, drawing on their insights into the interrelation and interdependence of plant and animal species. In his discussion of human ecology, Park (1936[1952]) begins with an explanation of the 'web of life', citing the familiar nursery rhyme, *The House that Jack Built*, as the logical prototype of long food chains, each link of which is dependent upon the other. Within the web of life, the active principle is the 'struggle for existence' in which the survivors find their 'niches' in the physical environment and in the division of labour among the different species.

If Park had been primarily interested in the natural environment for its own sake, he might have realised that human intervention in the form of urban development and industrial pollution artificially broke this chain, thereby upsetting the 'biotic balance'. In fact, he did acknowledge that commerce, in 'progresssively destroying the isolation upon which the ancient order of nature rested', has intensified the struggle for existence over an ever-widening area of the habitable world. But he believed that such changes had the capacity to give a new and often superior direction to the future course of events forcing adaptation, change and a new equilibrium.

Biological ecology was principally a source from which Park borrowed a series of principles which he applied to human populations and communities. In doing so, however, he notes that human ecology differs in several important respects from plant and animal ecology. First, humans are not so immediately dependent upon the physical environment, having been emancipated by the division of labour. Second, technology has allowed humans to remake their habitat and their world rather than to be constrained by it. Third, the structure of human communities is more than just the product of biologically determined factors; it is governed by cultural factors, notably an institutional structure rooted in custom and tradition. Human society, then, in contrast to the rest of nature, is organised on two levels: the biotic and the cultural.

This portrait of the nature–society relationship clearly contravenes many of the tenets of Catton and Dunlap's New Ecological Paradigm. It emphasises humans' *exceptional* characteristics (inventiveness, technical capability) rather than their commonality with other species. It gives priority to the influence of social and cultural factors (communication, division of labour) rather than biophysical, environmental determinants. Finally, it downplays the constraints imposed by nature by celebrating the human capacity to master it.

Park, his colleagues and students (notably McKenzie and Burgess) applied their principles of human ecology to the processes that create and reinforce urban spatial arrangements. They visualised the city as the product of three such processes: (1) concentration and deconcentration; (2) ecological specialisation; and (3) invasion and succession. The building blocks of the city were said to be 'natural areas' (slums, ghettoes, bohemias), the habitats of natural groups which were in accordance with these ecological processes. The city was depicted as a territorially based ecological system in which a constant Darwinian struggle over land use produced a continuous flux and redistribution of the urban population. Nowhere was this more evident than in the 'zone in transition', an area adjacent to the central business district which went from a coveted residential district to a blighted area characterised by low rent tenants, deviant activities and marginal businesses.

Much of the early criticism of human ecology rested not on its failure to explore the interdependence between the human environment and the natural environment but rather in what was perceived as its failure to adequately account for the role of human values in residential choice and movement. In the late 1940s, a sociocultural critique of mainstream human ecology briefly lit up the landscape of American sociology. Firey (1947) used the example of land use in central Boston to demonstrate that symbolism and sentiment were as important, if not more so, than standard ecological principles in accounting for the shape of the city. Similarly, Jonassen (1949) presented the history of settlement and relocation of

Norwegian immigrants to the New York City area as evidence that ethnic groups consciously *choose* a specific type of residential environment on the basis of values which they bring with them as a type of cultural baggage (in this case, the ideal included the sea, a harbour and mountains). Jonassen's research might have been the launching pad for a body of research on the origins of environmental perceptions (see for example Lynch's recent (1993) article on constructions of nature in Latin America) but the main thrust of his argument was rather to discredit the economic determinism which characterised the orthodox ecology of the day.

While cultural ecology, *per se*, never became dominant, it did force more traditional human ecologists to take greater account of social organisational and cultural variables. This was evident in O. D. Duncan's POET (Population–Organisation–Environment–Technology) model (1961) which was depicted as an 'ecological complex' in which: (1) each element is interrelated with the other three, and (2) a change in one can therefore affect each of the others. The POET model was a trailblazer in providing insight into the complex nature of ecological disruptions even if it failed to give sufficient weight to environmental constraints. For example, in a causal sequence suggested by Dunlap (1993: 722–3), an increase in population (P) can create a pressure for technological change (T) as well as increased urbanisation (O), leading to the creation of more pollution (E). While it was still rooted in orthodox human ecology, nevertheless, Duncan's POET model with its use of the human ecological complex at times 'came close to an embryonic form of environmental sociology' (Buttel and Humphrey, forthcoming, p.14).

The ecological basis of environmental destruction is probably best described in Catton and Dunlap's own 'three competing functions of the environment'[3] (see Figure 1).

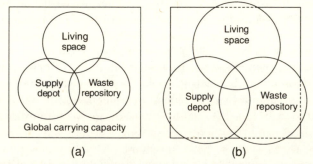

(a)

(b)

Figure 1 Competing functions of the environment: (a) circa 1900;
(b) current situation
Source: Dunlap 1993

Catton and Dunlap's model specifies three general functions which the environment serves for human beings: supply depot, living space and waste repository. Used as a *supply depot*, the environment is a source of renewable and non-renewable natural resources (air, water, forests, fossil fuels) that are essential for living. Overuse of these resources results in shortages or scarcities. *Living space* or habitat provides housing, transportation systems and other essentials of daily life. Overuse of this function results in overcrowding, congestion and the destruction of habitats for other species. With the *waste repository* function, the environment serves as a 'sink' for garbage (rubbish), sewage, industrial pollution and other byproducts. Exceeding the ability of ecosystems to absorb wastes results in health problems from toxic wastes and in ecosystem disruption.

Furthermore, each of these functions competes for space, often impinging upon the others. For example, placing a garbage landfill in a rural location near to a city both makes that site unsuitable as a living space and destroys the ability of the land to function as a supply depot for food. Similarly, urban sprawl reduces the amount of arable land which can be put into production while intensive logging threatens the living space of native peoples.

In recent years, the overlap, and therefore conflict, among these three competing functions of the environment has grown considerably. Newer problems such as global warming are said to stem from competition among all three functions simultaneously. Furthermore, conflicts between functions at the level of regional ecosystems now have implications for the global environment.

There are several very attractive features to Catton and Dunlap's competing functions of the environment model. First and foremost, it extends human ecology beyond an exclusive concern with living space – the central focus of urban ecology – to the environmentally relevant functions of supply and waste disposal. In addition, it incorporates a time dimension: both the absolute size and the area of overlap of these functions are said to have increased since the year 1900.

At the same time, there are problems with the model. As is the case with the urban ecology of Park and the Chicago School, there is no evidence of a human hand here. It says nothing about the social actions involved in these functions and how they are implicated in the overuse and abuse of environmental resources. Above all, there is no provision for changing either values or power relationships. The former is especially puzzling, since one would have thought that Catton and Dunlap would have attempted to link their ecological model to the new human ecology as emphasised in the HEP–NEP contrast. Finally, one cannot help comparing the longitudinal features of the Catton–Dunlap model to Beck's (1992) depiction of the transformation from an industrial to an industrial risk society. Both models recognise some of the same features: the increasing globalisation of environmental dangers, the rising prominence of output or waste-related elements as opposed to input or production-related ones. However, Beck's model is ultimately more exciting because it centrally incorporates the process of social definition. Beck's (1992: 24) criticism of environmental risk assessment, i.e. that 'it runs the risk of atrophying into a discussion of nature without people, without asking about matters of social and cultural significance', is equally applicable to Catton and Dunlap's competing functions of the environment.

Political economy explanation

Purveyors of the political economy explanation, by contrast, are quite clear about who they blame for the destruction of the environment: advanced industrial capitalism and its search for wealth, power and profit (Liazos 1989: 102). It follows from this that 'environmental issues are fundamentally social class issues' (Cable and Cable 1995: xi) in which the corporation and the state line up in opposition to ordinary citizens.

If the ecological explanation traces its pedigree to the human ecology of the Chicago School, the political economy approach draws its inspiration from the nineteenth-century writing of Karl Marx and Friedrich Engels. Marx and Engels were only marginally concerned with environmental degradation *per se* but their analysis of social structure and social change has become the starting point for several formidable contemporary theories of the environment.

Marx and Engels believed that social conflict between the two principal classes in society, i.e. capitalists and the proletariat (workers), not only alienates ordinary people from their jobs but it also leads to their estrangement from nature itself. Nowhere is this more evident than in 'capitalist agriculture' which puts a quick profit from the land ahead of the welfare of both humans and the soil. As the industrial revolution proceeded through the eighteenth and nineteenth centuries, rural workers were removed from the land and driven into crowded, polluted cities while the soil itself was drained of its vitality (Parsons 1977: 19). In short, a single factor, capitalism, was held responsible for a wide range of social ills from over-population and resource depletion to the alienation of people from the natural world with which they were once united. Marx and Engels saw the solution as the overthrow of the dominant system of production, capitalism, and the establishment in its place of a 'rational, humane, environmentally unalienated social order' (Lee 1980: 11).

Marx and Engels argue for the establishment of a new relationship between people and nature. However, it is not entirely clear what form such a relationship should take. In the work of the more mature Marx, this seems to follow a distinctly anthropocentric direction depicting humans as achieving mastery over nature, in no small part because of technological innovation and automation. By contrast, in Marx's early work the concept of the 'humanisation of nature' is proposed. This suggests that humans will develop a new understanding of and empathy with nature. A key question here is whether this new understanding would be used solely for human emancipation or whether it would take a more ecocentric form in which the powers and capacities of non-human species would be enhanced. In the former case, the humanisation of nature might, in fact, be deployed to eliminate species and organisms which threaten human health (Dickens 1992: 86). As Martell (1994: 152) observes, the texts of the early Marx are too complicated and contradictory on ecological concerns to be the basis for a fully fledged theory of environmental protection; it may be more useful to pursue this project through other sources or frameworks.

Contemporary Marxist theory emphasises not only the role of capitalists but also that of the state in fostering ecological destruction. Both elected politicians and bureaucratic administrators are depicted as being centrally committed to propping up the interests of capitalist investors and employers. While the incentive here is partly material, i.e. corporate campaign contributions, future job offers, public servants, politicians and capitalist producers are said to share an 'ethic' which accentuates capitalist accumulation and economic growth as the dual engines which drive progress. This, they argue, holds for all political levels from the global system to the local community.

Within environmental sociology, probably the most influential explanation of the relationship between capitalism, the state and the environment can be found in Alan Schnaiberg's book, *The Environment: From Surplus to Scarcity* (1980). Drawing on strands of both Marxist political economy and neo-Weberian sociology, Schnaiberg outlines the nature and genesis of the contradictory relations between economic expansion and environmental disruption.

Schnaiberg has depicted the political economy of environmental problems and policies as being organised within the structure of modern industrial society which he labels the *treadmill of production*. This refers to the inherent need of an economic system to continually yield a profit by creating consumer demand for new products, even where this means expanding the ecosystem to the point where it exceeds its physical limits to growth or its 'carrying capacity'. One particularly important tool in fuelling this demand is advertising, which convinces people to buy new products as much for reasons of lifestyle enhancement as for practical considerations.

Schnaiberg portrays the treadmill of production as a 'complex self-reinforcing mechanism' (Buttel and Humphrey, forthcoming) in which politicians respond to the environmental fall-out created by capital intensive economic growth by mandating policies which encourage yet further expansion. For example, resource shortages are handled not by reducing consumption or adopting a more modest lifestyle but by opening up new areas to exploitation.

Schnaiberg detects a *dialectic tension* which arises in advanced industrial societies as a consequence of the conflict between the treadmill of production and demands for environmental protection. He describes this as a clash between 'use values'; for example, the value of preserving existing unique species of plants and animals, and 'exchange values' which characterize the industrial use of natural resources. As environmental protection has emerged as a significant item on the policy agendas of governments, the state must increasingly balance its dual role as a facilitator of capital accumulation and economic growth and its role as environmental regulator and champion.

From time to time, the state finds it necessary to engage in a limited degree of environmental intervention in order to stop natural resources from being exploited with abandon and to enhance its legitimacy with the public. For example, in the progressive era of American politics in the late nineteenth and early twentieth centuries, the US government responded to uncontrolled logging, mining and hunting on wilderness lands by expanding its jurisdiction over the environment. Especially under the presidency of Theodore ('Teddy') Roosevelt, it created national forests, parks and wildlife sanctuaries, set limits and rules for the use of public lands and restricted the hunting of endangered species. It did so, however, as much out of a desire to increase industrial efficiency (Hays 1959), regulate competition and ensure a steady supply of resources (Modavi 1991) as it did from any sense of moral outrage. Similarly, the sudden emergence of toxic waste as a premier media issue in the early 1980s led to Congressional efforts in the United States to pass a new 'Superfund' law which would give the government statutory authority and the fiscal mechanisms to undertake clean-up operations without first having to legally identify the responsible parties. This was, Szasz (1994: 65) notes, not simply a matter of lawmakers addressing a newly recognised social need, but instead, 'one of those quintessential "time to make a new law" moments so characteristic of the American legislative process'.

Nevertheless, most governments remain wary of running the risk of slowing down the drive towards economic expansion or decelerating the treadmill of production (Novek and Kampen 1992). Caught in a contradictory position as both promoter of economic development and as environmental regulator, governments often engage in a process of 'environmental managerialism' (Redclift 1986), in which they attempt to legislate a limited degree of protection sufficient to deflect criticism but not significant enough to derail the engine of growth. By enacting environmental policies and procedures which are complex, ambiguous and open to exploitation by the forces of capital production and accumulation (Modavi 1991: 270), the state reaffirms its commitment to strategies for promoting economic development.

Other more stridently left-wing critiques have been even more unsparing in linking the dynamics of capitalist development to the rise in environmental destruction. David Harvey (1974), the Marxist geographer, accuses capitalist supremos of deliberately creating resource scarcities in order that prices may be kept high. Faber and O'Connor (1993) charge that the goal of capital restructuring in the 1980s and 1990s, which includes geographical relocation, plant closures and down-sizing, is to increase the exploitation of both the workers and nature; for example, by reducing spending on pollution control equipment. Cable and Cable refuse to rule out the possibility of insurrection in the United States if the grievances of grassroots environmental groups are ignored by capitalist economic institutions (1995: 121).

In his most recent writing with collaborator Kenneth Gould, Schnaiberg addresses the application of the treadmill of production to a Third World context. Ignoring the negative environmental impacts that the treadmill has produced in less developed regions, the leaders of Southern nations, in concert with the governments and corporations of the North, have sought to reproduce industrialisation as experienced by the First World. The primary mechanism for achieving this is the transfer of modern Western industrial techniques from North to South (Schnaiberg and Gould 1994: 167). However, as Redclift (1984) and others have noted, this transplant has become largely unsuccessful both in economic and environmental terms. Dependency on global markets has made economic development a risky venture for many Third World nations especially where these markets can easily be decimated by the appearance of new, low-cost alternatives elsewhere in the world. Furthermore, development schemes require an expensive infrastructure of roads, hydroelectric power dams, airports, etc., which must be paid for by borrowing heavily from Northern financial institutions. Such projects often fail to produce the expected level of economic growth while at the same time causing massive ecological damage in the form of flooding, rainforest destruction, soil erosion and pollution.

The political economy explanation has the advantage of locating present environmental problems in the inequities of humanly constructed political and economic systems rather than the abstract conflict of functions favoured by human ecologists. This brings it closer to the orbit of mainstream sociology than the more idiosyncratic approach advocated by Catton and Dunlap. Yet models such as Schnaiberg's are not without their own set of problems.

As the revelations of widespread environmental destruction throughout the former Soviet bloc and, more recently, China have indicated, it does not make sense to blame the ills of the planet exclusively on the logic of capitalism. A commitment to economic growth and development and the large-scale social structures which they sustain characterises most major post-Enlightenment ideologies including Marxism, liberalism and social democracy (Bahro 1984). Neo-Marxist writers have performed some remarkable contortions in trying to rationalise this contradiction. It is claimed, for example, that orthodox Communism had metamorphosised into a brand of state socialism which was fatally flawed by a combination of widespread bureaucratic inflexibility and corruption and an ill-conceived attempt to emulate capitalist economies. It is probably more accurate to say that both systems share a commitment to an unbridled industrialism for which the development ethic is vital.

Furthermore, unidimensional models such as Schnaiberg's seem better suited to addressing the classic cases of resource exploitation by primary producers or toxic dumping by petrochemical firms and other corporate polluters than they are in explaining the more recent transnational problems such as global warming or biodiversity loss. The latter are more multi-dimensional and complex, encompassing a wide range of different economic, social and political interests, ideologies and conflicts. While a central cast of capitalist actors (multinational corporations, the World Bank, the International Monetary Fund) is deeply implicated in the fate of the global environment, so too are a variety of supporting players – Faustian scientists in the biotechnology labs of California, maverick gold-miners in the Amazon rainforest, walrus poachers on the Bering Sea. There is a richly variegated political economy here but charting it requires a more flexible, historically sensitive approach than is possible with the more rigid, structurally centred models which have predominated thus far.

Political economy approaches can also be criticised for having adopted a monolithic view of the state as an environmental transgressor. It is more accurate to view politicians and civil servants as representing a variety of policy positions which are not always compatible. For example, Litfin (1993: 113–14) cites the prominent role of some national environmental agencies in constructing and promoting international agreements on pollution control as evidence of the positive entrepreneurial leadership role which some public servants can play in environmental politics. The campaign to save the North American whooping crane from extinction began in the early twentieth century with a few individual wildlife managers within the public service, notably Fred Bradshaw, Chief Game Guardian of Saskatchewan and head of the Provincial Museum of Natural History (Dunlap 1991). On occasion, differences may even emerge in full public view: the contrasting positions of President George Bush and his Secretary of the Environment at the June 1992 Earth Summit in Rio de Janeiro is one notable example (see pages 158–9).

Rise of environmental consciousness and movements

A second problem which is centrally addressed in the environmental sociology literature is that of why environmental consciousness and movements grew so dramatically from the early 1970s onwards in both Europe and America. Four main explanations have been put forward here: the reflection hypothesis; the post–materialism thesis; the new middle-class thesis; and the regulationist/political closure approach.

Reflection hypothesis

The reflection hypothesis[4] starts with the observation that environmental deterioration in Western industrial nations first began to climb after the Second World War, reaching its zenith by the late 1960s. The dramatic upswing after 1970 in environmental consciousness and concern is interpreted as a direct reaction to this worsening situation.

Circumstantial evidence for this position is provided by Dunlap and Scarce (1990), whose analysis of twenty years of polling data indicates that a majority of the American public has increasingly come to view a wide range of environmental problems as threatening both their personal health and the overall quality of the environment, and that this threat has increased markedly. Furthermore, this majority perceives environmental quality as deteriorating and likely to continue to do so.

More explicitly, Jehlicka (1992; cited in Martell 1994) argues that green

23

concern in Western Europe varies directly according to the seriousness of ecological conditions. Thus, in Southern Germany, Belgium, Luxembourg, the Netherlands, Northern France and Switzerland, where the pollution of rivers, forests and soils is most acute, environmental concern is highly developed. By contrast, in Britain and Scandinavia where environmental deterioration is less obvious, environmentalism is more moderate and absorbed into mainstream politics.

Other data, however, have not supported this reflection hypothesis. While environmental quality has been steadily deteriorating for much of this century, the public has ignored these developments for most of this period. When the Izaak Walton League, an established American conservation organisation, sponsored a national Clean Air Week in 1960 to try to acquaint the public with the existence of a national crisis, it encountered little popular interest or support (Trop and Roos 1971). Instead, perception of environmental problems may even be independent of the magnitude of the problems themselves. For example, concern about air pollution arose in the United States in the late 1960s at the same time as the levels of a number of common air pollutants were found to have declined in a broad sample of urban areas (Albrecht and Mauss 1975: 573). This suggests that public concern is at least partially independent of actual environmental deterioration and is shaped by other considerations; for example, the extent of mass media coverage.

Furthermore, most of the modern environmental problems, particularly second generation problems such as acid rain, global warming, ozone depletion and toxic contamination, are likely to be invisible to the naked eye except in the most extreme cases (Cylke 1993: 26). As a result, the public perception that environmental problems have reached 'crisis' proportions does not necessarily reflect the reality of actual problems but rather the particular view of scientific experts, environmentalists and the media.

Post-materialism thesis

A second explanation locates environmental concern as part of a more extensive shift in values among certain segments of Western societies. This approach has as its touchstone Inglehart's (1971, 1977, 1990) 'post-materialism thesis'.

Inglehart's interpretation is derived from the 'hierarchy of needs' proposed by the humanistic social psychologist, Abraham Maslow (1954). Inglehart proposes that the economic worries experienced by an older generation during the Great Depression and the two World Wars had little meaning for the post-Second World War 'baby boom' generation which had the financial security to allow them instead to address their non-material needs for belonging and individual fulfilment. This cohort was less interested in promoting economic growth and progress than in furthering post-materialist values such as a concern for ideas, the pursuit of personal growth, autonomy in decision-making and improving the quality of the physical environment. Significantly, post-materialism was not simply a life cycle phenomenon, fading out of existence when the post-war generation settled down and started families of their own, but a lasting value change (Inglehart 1977).

Inglehart's thesis was first substantively linked to environmentalism by the British sociologist, Stephen Cotgrove (1982). Cotgrove, using a modified form of Inglehart's post-material values scale, found that environmentalists were the most likely of any of the four publics that he surveyed (the others being industrialists, trade union officials and nature conservationists) to qualify as post-materialists, attaching high priority to non-material goals such as giving people more input into important government decisions and progressing towards a less impersonal, more humane society. Cotgrove concluded that what distinguishes members of new environmental groups (in this case the Friends of the Earth) is not just their heightened environmental awareness but a commitment to a set of alternative, post-materialist values. Furthermore, Cotgrove's analysis suggested that a person's values (anti-economic, post-material) and attitudes (anti-industrial, perhaps anti-science) are the key predictors of heightened environmental concern within other constituencies as well, notably conservationists who worry about environmental damage and threats to nature.

Thus, in contrast to the reflection hypothesis, the growth of environmental consciousness and concern is not seen as being directly related to the actual extent to which the environment has deteriorated. As Cotgrove notes, the 'objective facts' about pollution and environmental damage and shortages do not and cannot exist in some kind of cognitive and moral vacuum, but rather arise from a moral debate over the nature of the good society which cannot 'easily be settled by [an] appeal to facts and rational argument' (1982: 32).

The post-materialist thesis has recently been challenged by Brechin and Kempton (1994), who demonstrate that public environmental concern is not just restricted to advanced industrial countries but exists on a global scale. They present two types of evidence to support this: widespread grassroots environmental activism and a pair of cross-national opinion surveys. Brechin and Kempton's survey data analysis reveals higher percentages of respondents in some Third World nations (India, Mexico, Uruguay) willing to pay higher prices and taxes in order to protect the environment than is the case in some more industrialised nations such as Finland and Japan. Environmentalism, they conclude, should not be viewed as a product of a post-materialist shift in values but rather appears to be a more complicated phenomenon, emerging from multiple sources in richer and poorer nations alike.

The problem is that it is never made clear where these post-materialist values originate. It may be surmised that they are a function of interests; for example, industrialists can be expected to oppose an ideal society which, among other things, accepts a no-growth philosophy or one which is predominantly socialist. It is not as easy to figure out where post-materialists, including environmentalists, get their values. Cotgrove promises to answer this question in the conclusion to his second chapter but then only tells us that a commitment to non-material values is forged in adolescence, is part of a long-term drift away from any strong allegiance to the culture of business and is more likely to occur in homes where the parents have already embraced post-material values. He does, however, observe that environmentalism is an expression of the interests of a new middle-class fraction who dissent from traditional paradigms which emphasise pro-business values. This is the basis for the third sociological explanation for the growth of environmental consciousness and concern: the new middle-class thesis.

New middle-class thesis

The new middle-class thesis is a companion to the post-materialism thesis but it puts a greater emphasis on the social location of those who adopt an environmentalist ethic. According to this view, environmentalists are drawn disproportionately from that segment of society which has been termed 'social and cultural specialists' – teachers, social workers, journalists, artists and professors who work in creative and/or public service-oriented jobs (Cotgrove and Duff 1981; Kriesi 1989).

It is not entirely clear why this occupational segment should be more inclined to produce environmentalists with post-material values as against other sections of the middle classes. One possible explanation lies in the nature of their involvement and interaction with their clients. By virtue of their positions, they are socially situated so as to witness first hand the victimisation of the powerless by the heralds of industrial progress. For example, doctors staffing a community health clinic are strategically located to witness the adverse effects on schoolchildren of elevated lead levels in the soil of neighbourhoods built around polluting, inner city factories. As a result, they tend to become personally involved in environmental problems, even to the point of becoming advocates for their patients' interests. Alternatively, it may simply be that those who enter professions which have a significant creative or social welfare component may choose these deliberately, guided by an already existing post-materialist value orientation. By contrast, those who are more interested in technical or financial goals choose to work in banks, engineering firms, public works departments, etc. In reality, it is probably some combination of these two explanations which is operative here.

A useful comparison may be made to the extensive involvement of Catholic religious orders in movements for social change in Latin America, the Philippines and other Third World nations. Initially guided by certain altruistic values, it is only when missionaries from Ireland and other European nations directly encounter the often violent realities of life among the 'shirtless ones' in despotic regimes that they adopt an explicitly activist and often radical perspective. Similarly, members of the new middle class may enter their jobs possessing certain inclinations but it is the fact of being in the firing line of environmental injustice that pushes them towards a more explicit ecological consciousness.

An alternative explanation which has been associated with Peter Berger (1986) suggests that this new knowledge class is not so much altruistic as intensely cognisant of its own interests. Since they are the ones most likely to enjoy the positive organisational fruits of NSM activism – jobs in universities, government departments, regulatory agencies and pressure groups, research grants, conference travel, etc. – it is not suprising that members of the new middle class make up the bulk of the constituency of support for environmentalism, feminism, anti-nuclearism, etc.

Steinmetz (1994: 183–4) has identified two major difficulties in attempting to explain NSMs, such as environmentalism, in terms of the rise of a new middle class.

First, he notes that recent research has indicated that the social composition of NSMs is more diverse than the class explanation has acknowledged. For example, he cites evidence from public opinion polls and voting patterns in Germany in the late 1980s which indicates that the distribution of support for *Die Grunen* (Greens) was, in fact, flattening out. This is consistent with recent research in the United States on the environmental justice movement which reports a rising presence in environmental protests by members of disadvantaged groups (see Chapter 6). Steinmetz cites Beck's observation that 'need is hierarchical, smog is democratic' to illustrate that in the contemporary 'risk society' we are all centrally affected by environmental problems, a fact which sooner or later will lead to increased environmental consciousness across class lines.

Second, he observes that even if the thesis that the new middle class is over-represented could be supported empirically, this may indicate that this group is simply better able to perceive and mobilise against problems such as environmental deterioration than are the equally concerned but less positively resourced lower classes. As it happens, segments of the middle class were similarly over-represented in many 'old' social movements (Bagguley 1992) – a further indication that they possess resources (flexible time, leadership skills, etc.) which allow them to participate more intensely.

Regulationist/political closure approach

Finally, there have been attempts to account for the rise of environmental consciousness and action by identifying tensions in the political systems of some Western European nations.

From this perspective, the New Social Movements are said to have arisen as a defensive reaction against the intrusion of the state into everyday life of ordinary citizens – what Habermas (1987) terms the 'colonisation of the life world'. While this generally fits better as an explanation for the growth of social movements organised around alternative sexual identities and lifestyles, it can also be seen as having some relevance to the environmental sphere.

It can be argued, for example, as does Beck (1992), that the proliferation of new chemical and nuclear, and, most recently, biogenetic technologies, has brought a host of new risks into the daily lives of modern citizens. Governments have sometimes been the architects of these risks; at other times the henchmen of those who are the risk creators. Halfmann and Japp (1993: 438) argue that modern social movements such as the environmental movement choose as their targets risks which appear to represent the ultimate threat to our 'life chances' because they seem to be uncontrollable and irreversible: nuclear power plants, deranged ecosystems, the arms race and biotechnology.

Another plane of this structural explanation casts the rise of environmentalism in the context of 'neo-corporatism'. Corporatist-type political arrangements exist when the state joins in partnership with private industry and sometimes big labour unions to circumvent formal democratic procedures and make key political and economic decisions behind closed doors. Frequently, this form of circumscribed decision-making can result in damage to the environment, especially since corporatism is premised on sustained economic growth and high levels of employment (Scott 1990: 146). For example, the 1992 Summer Olympic Games in Barcelona, which were organised under the aegis of a public limited company which conjoined private capital with regional and state governments, resulted in environmental damage at a number of locations including the natural park of Collserola, the last remaining habitat for a number of plant and animal species (Tabara 1992; Tabara and Hannigan 1993).

It is argued that the political closure imposed by corporatist arrangements has precipitated new forms of ecological protest. NSMs are said to have arisen outside mainstream politics in civil society in order to address grievances and themes (including ecological destruction) which have been systematically marginalised by the corporatist state. Such issues have been officially excluded because they are of no significance or challenge to the interests of the major parties in the corporatist partnership (Hirsch and Roth 1986). Scott (1990) observes that it is in nations in which political debate has been stifled under a real or apparent consensus and decision-making dominated by a small group of 'social partners' (i.e. Austria, Germany, Sweden) that ecological movements, notably the Greens, have been most active in the political sphere.

In West Germany, for example, bureaucratic policy-makers had, by the 1970s, increasingly begun to avoid parliamentary institutions, preferring to make key decisions in concert with industry representatives behind closed doors. The rise of *Die Grünen* can thus be interpreted as an attempt to re-establish the democratic political link between the state and the citizenry, first through the formation of extra-parliamentary citizen initiative groups, and later by re-entering parliaments in the form of alternative parties with the goal of helping to restore parliamentary legitimacy (Hager 1993).

While these regulationist/political closure explanations have the advantage of placing the rise of environmentalism in a wider historical and cultural context, they tell us more about the structural source and channelling of grievances which are held by environmental activists than they do about any individual motivation to embrace a 'green' view of the world (Steinmetz 1994: 195–6). Furthermore, while it is possible to understand why in some European nations the centre of gravity of environmental discourse was to be found in ecological movements rather than in politics, it is less evident how environmentally related grievances came to be constructed into full-blown claims within these newly emergent green networks. This is especially relevant since these decentralised NSM groups tend to work out their new collective meanings and identities in a pre-political or private context rather than in the full glare of politics and public policy-making (Melucci 1989).

CONCLUSION

While each of the approaches discussed thus far has its merits, none is able to adequately account for the manner in which environmental problems are defined, articulated and acted upon by social actors. For example, why did environmentalism remain in a relative state of abeyance for a half century from 1920 to 1970? Why have global environmental problems such as ozone depletion, global warming and biodiversity loss displaced more local problems such as groundwater pollution and urban sewage disposal as priorities for governments, the media and the mainstream environmental movement? Why do some arcane scientific findings become the basis for high-profile environmental problems while others languish in obscurity? These are the types of questions which are addressed by the perspective that I introduce in the next chapter: the social constructionist approach to environmentalism and the environment. A social constructionist perspective on the environment has several advantages over other theoretical approaches.

In contrast to much of the existing sociological literature on the environment, social constructionism does not uncritically accept the existence of an environmental crisis brought on by unchecked population growth, over-production, dangerous new technologies, etc. Instead, it focuses on the social, political and cultural processes by which environmental conditions are defined as being unacceptably risky and therefore actionable. As Thompson (1991) has noted, environmental debates reflect the existence not just of an absence of certainty (e.g. about energy futures, the extent of the hazardous waste problem, the health effects of low level radiation) but rather the existence of *contradictory certainties*: severely divergent and mutually irreconcilable sets of convictions both about the environmental problems we face and the solutions that are available to us.

It is important to note, however, that environmental risks and problems as socially constructed entities need not undercut legitimate claims about the condition of the environment, thereby denying them an objective reality (Short 1992; Ungar 1992). As Yearley (1992: 186) observes, demonstrating that a problem has been socially constructed is not to undermine or debunk it, since 'both valid and invalid social problem claims have to be constructed'. Similarly, social constructionism as it is conceptualised here does not deny the independent causal powers of nature but rather asserts that the rank ordering of these problems by social actors does not always directly correspond to actual need. To a considerable extent, this reflects the political nature of agenda-setting. As Bird (1987: 260) has argued, understanding how environmental problems have been socially and politically negotiated gives them 'enormous normative weight'.

Second, much of the manufacturing of environmental problems is carried out in arenas that are populated by communities of specialists: scientists, engineers, lawyers, medical doctors, government officials, corporate managers, political operatives, etc., rather than in full view of the general public[5] (Hilgartner 1992: 51–2). As a result, research perspectives which focus exclusively on public discourse fail to fully capture the details of environmental agenda-setting and policy-making. A social constructionist approach, by contrast, recognises the extent to which environmental problems and solutions are end-products of a dynamic social process of definition, negotiation and legitimation both in public and private settings.

Third, a social constructionist approach grounds the study of environmental matters in a distinctly sociological paradigm. By contrast, much of what has heretofore fallen under the label of 'environmental sociology' arises from an extra-disciplinary discourse which demands that the analyst subscribe to a new set of ecological values. As Lundquist (1978: 89–90) has argued for the case of environmental political science, the primary goal of studying environmental problems should be to push forward the frontiers of the discipline rather than to secure an ecologically sound 'Futuropia'. That is not to say that environmental sociologists should deny the seriousness of the threats faced by our planet; nor are they advised to embrace the growth-centred ideology which characterised mainstream sociology in the past. Rather, we should deliberately adopt the agnostic stance required by a constructionist approach (Yearley 1992: 186) in order to optimally assess how environmental knowledge, risks and problems are socially assembled.

2

SOCIAL CONSTRUCTION OF ENVIRONMENTAL PROBLEMS

The constructionist approach to environmental problems has multiple origins, but it is best understood by returning to the early 1970s when conventional explanations for the existence of social problems were seriously challenged for the first time.

CONSTRUCTING SOCIAL PROBLEMS

Nearly a quarter of a century ago, the sociology of social problems first began to experience a major paradigmal conflict with the appearance of a seminal article by Malcolm Spector and John Kitsuse (1973) entitled 'Social problems: a reformulation'. Here, and in a subsequent book (1977), Spector and Kitsuse challenged the 'structural functional' approach to social problems which had theretofore dominated the field. Functionalism, as exemplified by the work of Merton and Nisbet (1971), assumed the existence of social problems (crime, divorce, mental illness) which were the direct products of readily identifiable, distinctive and visible objective conditions. Sociologists were regarded as experts who employ scientific methods to locate and analyse these moral violations and advise policy-makers on how best to cope. In addition, the sociologist's role was to bring to lay audiences an awareness and understanding of worrisome conditions, especially where these were not readily evident (Gusfield 1984: 39).

Spector and Kitsuse argued that social problems are not static conditions but rather 'sequences of events' which develop on the basis of collective definitions. Accordingly, they defined social problems as 'the activities of groups making assertions of grievances and claims to organizations, agencies and institutions about some putative conditions' (1973: 146). From this point of view, the process of claims-making is treated as more important than the task of assessing whether these claims are truly valid or not. For example, rather than document a rising crime rate, the social problems analyst is urged to focus on how this problem is 'generated and sustained by the activities of complaining groups and institutional responses to them' (1973: 158).

Since 1973, social constructionism has increasingly moved towards the core of social problems theorising, generating a critical mass of theoretical and empirical contributions (see especially Best 1989a; Gusfield 1981; Holstein and Miller 1993; Loseke 1992; Schneider 1985; Schneider and Kitsuse 1984). Constructionism has gained currency in other scholarly specialty areas as well, notably science and technology (Knorr-Cetina 1983; Latour and Woolgar 1986; Pinch and Bijker 1987), gender relations (Laws and Schwartz 1977; Mackie 1987) and media studies (Altheide 1976; Fishman 1980; Schlesinger 1978). In each case, what a constructionist analysis has in common is a concern with how people assign meaning to their world (Best 1989b: 252).

Controversies

While the social constructionist approach has both transformed and re-vitalised the sociology of social problems it has not been immune from controversy.

One central debate in constructionist analysis concerns the relative or contingent nature of social problems. Conventional social problems analysts have warned that constructionism runs the risk of denying the harmful existence of serious 'real-life' problems – a charge that is similarly levelled at the social constructionist perspective on the environment. It does so, they claim, by making these conditions subject to the vagaries of social definition. At the opposite end of the spectrum, other critics take social constructionists to task for failing to entirely abandon the objective residues of functionalist theory. Most notably, Woolgar and Pawluch (1985) charge constructionism with engaging in the strategy of 'ontological gerrymandering'. By this they mean that constructionist authors continue to arbitrarily identify problematic conditions or behaviour worthy of study at the same time as relativising the definitions and claims made about them. Typically, a condition such as heroin use is treated as objectively real and constant over time while the social evaluation of this condition as problematic or not varies from era to era. This is internally inconsistent, Woolgar and Pawluch argue, since it distinguishes between a set of fixed conditions as identified by the social problems analyst and a set of changing, contextual conditions as proposed by social problems participants.

Social constructionists have responded to this charge of ontological gerrymandering in several ways. 'Strict constructionists' maintain that we must be ever more vigilant in making any assertions at all about social conditions. Instead, they favour turning to the ethnomethodological perspective in order to discover new ways of writing social constructionist texts which are centred entirely in the interpretations and practices of participants in social problem construction. By contrast, contextual constructionists argue that any claim can be evaluated on the basis of hard evidence such as official statistics or public opinion polls (Best 1989b: 247), even if these are in themselves social constructions. For example, Best (1993: 139) suggests that the social problems analyst can reasonably doubt claims that Satanists sacrifice 60,000 victims annually while accepting figures provided by the Centers for Disease Control for the numbers of American Aids victims. In particular, the researcher is encouraged to consider the historical context within which the social problems claim has been formulated in order to explain the emergence and assess the validity of this claim (Rafter 1992).

Social constructionism has also been divided over the utility of a 'natural history' of social problems. Spector and Kitsuse's original formulation contained a four-stage natural history model in which claims-making moves from attempts to transform private troubles into public issues through official recognition, dissatisfaction with the manner in which bureaucratic organizations are dealing with the imputed conditions, and finally, the development of alternative, parallel or counter-institutions to seek radically new solutions for perceived problems. However, as Schneider (1985: 225) has noted, this kind of natural history model encourages us to overstate the extent to which types of activities occur only at certain stages in the claims-making process. In fact, the stages proposed by Spector and Kitsuse tend to overlap (Wiener 1981) rather than follow an orderly succession. Hilgartner and Bosk (1988) have argued that it is now time to move beyond natural history models and propose an alternative 'ecological' model in which a population of potential social problems compete for societal attention within public arenas.

Constructionism as an analytic tool

Best (1989b: 250) has noted that constructionism is not only helpful as a theoretical stance but it can also be useful as an analytic tool. In this regard, he suggests three primary foci for studying social problems from a social constructionist perspective: the claims themselves; the claims-makers; and the claims-making process.

Nature of claims

As initially conceptualised by Spector and Kitsuse, claims were complaints about social conditions which members of a group perceived to be offensive and undesirable. According to Best (1989b: 250), there are several key questions to be considered when analysing the content of a claim: What is being said about the problem? How is the problem being typified? What is the rhetoric of claims–making and how are claims presented so as to persuade their audiences? Of these, it is the third question which has generated the most interest among contemporary social problems analysts.

Using the example of the 'missing children', i.e. runaways, child-snatch-ings and abductions by strangers, Best (1987) analyses the content of social problems claims by focusing on the 'rhetoric' of claims-making. Rhetoric involves the deliberate use of language in order to persuade. Rhetorical statements contain three principal components or categories of statements: grounds, warrants and conclusions.

Grounds or data furnish the basic facts which shape the ensuing claims-making discourse. There are three main types of grounds statements: definitions, examples and numeric estimates. Definitions set the boundaries or domain of the problem and give it an orientation; that is, a guide to how we interpret it. Examples make it easier for public bodies to identify with the people affected by the problem, especially where they are perceived as helpless victims. Atrocity tales are one especially effective type of example. By estimating the magnitude of the problem, claims-makers establish its importance, its potential for growth and its range (often of 'epidemic' proportions).

Warrants are justifications for demanding that action be taken. These can include presenting the victim as blameless or innocent, emphasising links with the historical past or linking the claims to basic rights and freedoms. For example, in analysing the professional literature on 'elder abuse', Baumann (1989) identified six primary warrants: (1) the elderly are dependent; (2) the elderly are vulnerable; (3) abuse is life-threatening; (4) the elderly are incompetent; (5) ageing stresses families; (6) elder abuse often indicates other family problems.

Conclusions spell out the action which is needed to alleviate or eradicate a social problem. This frequently entails the formulation of new social control policies by existing bureaucratic institutions or the creation of new agencies to carry out these policies.

Best further proposes two rhetorical themes or tactics which vary according to the nature of the target audience. The *rhetoric of rectitude* (values or morality require that a problem receive attention) is most effective early on in a claims-making campaign when audiences are more polarised, activists are less experienced and the primary demand is for a problem to be viewed in a new way. By contrast, the *rhetoric of rationality* (ratifying a claim will earn the audience some type of concrete benefits) works best at the later stages of social problems construction when claims-makers are more sophisticated, the primary demand is for detailed policy agendas and audiences are more persuadable. Rafter (1992: 27) has added another rhetorical tactic to Best's list: that of archetype formation. *Archetypes* are the templates from which stereotypes are minted and therefore possess considerable persuasive power as part of a claims-making campaign.

A further set of rhetorical strategies in claims-making has been proposed by Ibarra and Kitsuse (1993) who outline a variety of rhetorical idioms, motifs and claims-making styles.[1]

Rhetorical idioms are image clusters which endow claims with moral significance. They include a 'rhetoric of loss' (of innocence, nature, culture, etc.); a 'rhetoric of unreason' which invokes images of manipulation and conspiracy; a 'rhetoric of calamity' (in a world full of deteriorating conditions, epidemic proportions are claimed for a few; for example, Aids or the greenhouse effect); a 'rhetoric of entitlement' (justice and fair play demand that the condition, or as Ibarra and Kitsuse term it, the 'condition-category', be redressed), and the 'rhetoric of endangerment' (condition-categories pose intolerable risks to one's health or safety).

Rhetorical motifs are recurrent metaphors and other figures of speech (Aids as a 'plague', the depletion of the ozone layer as a 'ticking time bomb') which highlight some aspect of a social problem and imbue it with a moral significance. Some motifs refer to moral agents, others to practices and still others to magnitudes (Ibarra and Kitsuse 1993: 47).

Claims-making styles refer to the fashioning of a claim so that it is in sync with the intended audience (public bodies, bureaucrats, etc.). Examples of claims-making styles include a scientific style, a comic style, a theatrical style, a civic style, a legalistic style and a subcultural style. Claims-makers must match the right style to the right situation and audience.

Claims-makers

In looking at the identity of claims-makers, Best (1989b: 250) advises that we pose a number of questions. Are claims-makers affiliated to specific organisations, social movements, professions or interest groups? Do they represent their own interests or those of third parties? Are they experienced or novices (as we have seen, this can influence the choice of rhetorical tactics)?

Many studies which have been undertaken in the social constructionist mode have pointed to the important role played by medical professionals and scientists in constructing social problems claims. Others have noted the importance of policy or issue entrepreneurs – politicians, public interest law firms, civil servants whose careers are dependent upon creating new opportunities, programmes and sources of funding, etc. Claims-makers may also reside in the mass media, especially since the manufacture of news depends upon journalists, editors and producers constantly finding new trends, fashions and issues. The cast of claims-makers who combine to promote a social problem can sometimes be quite diverse. For example, Kitsuse *et al.* (1984) identify three main categories of claims-makers in the identification of the *kikokushijo* problem in Japan, i.e. the educational disadvantage of Japanese schoolchildren whose parents have taken them abroad as part of a corporate or diplomatic posting: officials in prestigious and influential government agencies; informally organised groups of diplomatic and corporate wives; and the *'meta'* – a support group of young adults who have been victims of the *kikokushijo* experience.

Claims-making process

Wiener (1981) has depicted the collective definition of social problems as a continually ricocheting interaction among three sub-processes: *animating the problem* (establishing turf rights, developing constituencies, funnelling advice and imparting skills and information); *legitimating the problem* (borrowing expertise and prestige, redefining its scope, e.g. from a moral to a legal question, building respectability, maintaining a separate identity); and *demonstrating the problem* (competing for attention, combining for strength, i.e. forging alliances with other claims-makers, selecting supportive data, convincing opposing ideologists, enlarging the bounds of responsibility). These are overlapping rather than sequential processes which together result in a public arena being built around a social problem.

Hilgartner and Bosk (1988) have identified these arenas of public discourse as the prime location for the evaluation of social problem definitions. However, rather than examining the stages of problem development, they propose a model which stresses the competition among potential social problems for attention, legitimacy and societal resources. Claims-makers or 'operatives' are said to deliberately adapt their social problem claims to fit their target environments; for instance, by packaging their claims in a novel, dramatic and succinct form or by framing claims in politically acceptable rhetoric.

Best (1989b: 251) poses a number of useful questions about the claims-making process. Whom did the claims-makers address? Were other claims-makers presenting rival claims? What concerns and interests did the claims-makers' audience bring to the issue, and how did these shape the audience's responses to the claims? How did the nature of the claims or the identity of the claims-makers affect the audience's response?

SOCIAL CONSTRUCTIONISM AND THE ENVIRONMENT

As we have already observed, environmental problems are similar in many ways to social problems in general. There are, however, a few notable differences. While social problems frequently cross over from a medical discourse to the arenas of public discourse and action (Rittenhouse 1991: 412), they nevertheless derive much of their rhetorical power from moral rather than factual argument. For example, the recent elevation of 'date rape' to social problem status probably owes more to the changing moral landscape of gender relations than it does to scientific evidence suggesting a sudden upswing in the incidence of this condition. By contrast, environmental problems such as pesticide poisoning or global warming, while morally charged, are tied more directly to scientific findings and claims (Yearley 1992: 117).

Furthermore, although they are traceable to human agents, environmental problems have a more imposing physical basis than social problems, which are more rooted in personal troubles that become converted into public issues (Mills 1959).

While the environment has never been of major interest to social problems researchers, it has received some limited attention, primarily in undergraduate social problems texts.[2] Almost uniformly, environmental problems are depicted here as real, identifiable and intrinsically harmful. In one text from the early 1980s, the authors even entitle a section of their environmental chapter 'Objective dimensions of the problem' and go on to discuss the 'scope' of air, water and pesticide pollution (Wright and Weiss 1980). A second social problems text published in the same year structures its chapter on 'The environment' around four questions: What is the cause of the environmental crisis? What are the long-term effects of pollution? How can we deal with growing shortages of natural resources? Is an ecological disaster avoidable? The authors, respected American sociologists James Coleman and Donald Cressey, briefly illustrate the constructionist definition of a social problem by noting that 'pollution did not become a social problem until environmental activists were able to convince others to show concern about conditions that had actually existed for some time'. However, they then undercut this by asking: 'If thousands of people did not know they were being poisoned by radiation leaking from a nuclear power plant, wouldn't radiation pollution still be a social problem?' (1980: 3–4).

One of the few social problems texts to deliberately attempt a constructionist perspective is Armand Mauss's 1975 book, *Social Problems as Social Movements*. In a chapter co-authored with environmental sociologist Stan Albrecht, Mauss deals briefly with such topics as 'cultural definitions of the environment' and 'political and scientific publics and interests', although much of the rest of the chapter is devoted to a history of the American environmental movement.

While contemporary social constructionist authors sometimes use examples relating to environmental issues and problems (cf. Ibarra and Kitsuse 1993), none of the major edited social problems readers undertaken from this perspective include an article on environmental problems. Instead, the impetus for a social constructionist perspective on the environment has come largely from within environmental sociology itself.[3]

Several prominent contributors to this field have recently called for the development of a constructionist model which could help guide future research into the creation and legitimation of environmental knowledge and risk. Freudenburg and Pastor have outlined a conceptual perspective which focuses on the framing of risk debates by institutional actors. Their 'social construction of risk conflicts' is justified on the basis that it 'brings risk to sociology rather than the converse, highlighting instead of hiding the political and discursive struggles embedded in technological risks' (1992: 398). In a similar vein, Buttel and Taylor have argued that environmental sociology must give more attention to the social construction of environmental knowledge. The global construction of environmental issues is, they contend, 'as much or more a matter of the social construction and politics of knowledge production as it is a straightforward reflection of biophysical reality' (1992: 214). A third author who explicitly draws on a constructionist model is Stella Capek. Capek (1993) uses a variety of sources from the social movement and social problems literatures including Best, Gusfield and Spector and Kitsuse to explain the emergence of an 'environmental justice' frame and its mobilising power in community struggles over toxic contamination in the Southern United States. Finally, Steven Yearley (1992), who also draws on Spector and Kitsuse, has examined the 'green case', i.e. the rise of environmental awareness and action over the last two decades, from the perspective of moral entrepreneurship and claims-making.

Empirically, the central role of claims-making activities in shaping environmental agendas, assessments and policies has been examined in analyses of chemical contamination (Aronoff and Gunter 1992), global climatic change (Hart and Victor 1993; Ungar 1992), media coverage of environmental issues and conflicts (Burgess and Harrison 1993; Hansen 1991; Mazur and Lee 1993; Schoenfeld *et al.* 1979) and risk and safety issues (Spencer and Triche 1994; Stallings 1990).

If the social constructionist perspective is compatible with any other approach to the environment, it is probably that of political economy. As will be variously noted in the following pages, the way in which environmental knowledge and risk are conceptualised and the relative success of these constructions are constrained by and channelled through existing structures of economic and political power. Yet political economy alone is not sufficient to explain the career paths of environmental problems. Perception is more than simply a function of power; it depends on a host of other factors which relate to culture and knowledge. It is important, therefore, to address the question posed by Benton and Redclift:

> What are the processes of communication, discursive processing, normative orientation, 'moral entrepreneurship' by which the antagonisms of the environmental debate get formed and transformed?
>
> (1994: 9)

Key tasks/processes

In defining environmental problems, bringing them to society's attention and provoking action, claims-makers must engage in a variety of activities. Some of these are centrally concerned with the collective definition of potential problems, others with the collective action necessary to ameliorate them (Cracknell 1993: 4). This is not to say that elements of definition and action do not interweave constantly. Nevertheless, environmental problems do follow a certain temporal order of development as they progress from initial discovery to policy implementation.

In this section of the chapter, I identify three central tasks which characterise the construction of environmental problems. In doing so, I draw upon two prior models: Carolyn Wiener's (1981) three processes through which a public arena is built around a social problem, and William Solesbury's (1976) three tasks which are necessary for an environmental issue to originate, develop and grow powerful within the political system.

As already noted earlier in this chapter, in her book *The Politics of Alcoholism*, Wiener depicted the collective definition of social problems as a continuing ricocheting interaction among three processes: animating, legitimising and demonstrating the problem. These are presented as over-lapping rather than sequential processes; that is, they interact with one another rather than operate independently.

Solesbury's scheme is more concerned with the political fate of environmental concerns. He notes the 'continuing change in the agenda of environmental issues' which may be partly accounted for by changes in the state of the environment itself (see Ungar 1992) and partly through changing public views as to which issues are important and which are not. All environmental issues, he states, must pass three separate tests: commanding attention, claiming legitimacy and invoking action. Like Wiener, Solesbury points out that these tasks may be pursued simultaneously in no particular order (Cracknell 1993: 5), although it would presumably be difficult to invoke policy changes before the problem is recognised and legitimised.

In considering the social construction of environmental problems, it is possible to identify three key tasks: assembling, presenting and contesting claims (Table 1, page 42).

Assembling environmental claims

The task of assembling environmental claims concerns the initial discovery and elaboration of an incipient problem. At this stage, it is necessary to engage in a variety of specific activities: naming the problem, distinguishing it from other similar or more encompassing problems, determining the scientific, technical, moral or legal basis of the claim, and gauging who is responsible for taking ameliorative action.

Environmental problems frequently originate in the realm of science. One reason for this is that ordinary people have neither the expertise nor the resources to find new problems. For example, knowledge about the ozone layer is not tied to our everyday experience; it is available only through the use of high technology probes into the atmosphere above the polar regions (Yearley 1992: 116).

Some problems, however, do relate more closely to our life experiences. Concern over toxic wastes frequently begin with local citizens who come to draw a causal link between seeping dump sites and a perceived increase in the neighbourhood incidence of leukaemia, miscarriages, birth defects and other ailments. This is what occurred in Niagara Falls, New York State, where Lois Gibbs and her neighbours were the first to associate their

Table 1   Key tasks in constructing environmental problems

| | Task | | |
	Assembling	Presenting	Contesting
Primary activities	• discovering the problem • naming the problem • determining the basis of the claim • establishing parameters	• commanding attention • legitimating the claim	• invoking action • mobilising support • defending ownership
Central forum	• Science	• Mass media	• Politics
Predominant 'layer of proof'	• Scientific	• Moral	• Legal
Predominant scientific role(s)[1]	• Trend spotter • Theory tester	• Communicator	• Applied policy analyst
Potential pitfalls	• lack of clarity • ambiguity • conflicting scientific evidence	• low visibility • declining novelty	• co-optation • issue fatigue • countervailing claims
Strategies for success	• creating an experiential focus • streamlining knowledge claims • scientific division of labour	• linkage to popular issues and causes • use of dramatic verbal and visual imagery • rhetorical tactics and strategies	• networking • developing technical expertise • opening policy windows

Note: 1 See Susskind (1994)

health-related problems with the chemical wastes buried thirty years before in the abandoned Love Canal.

Those whose jobs or recreational pursuits bring them into close contact with nature on a daily basis (farmers, anglers, wildlife officers) may als the initial source of claims because they pick up early environmental warning signals such as reproductive problems in livestock or mutations in fish. Acid rain was first launched as a contemporary environmental problem when a fisheries inspector in a remote area of Sweden telephoned re-searcher Svante Oden with the observation that there appeared to be a link between a rising incidence of fish deaths and an elevation in the acidity of lakes and rivers in the area.

Practical knowledge about the environment often originates from the everyday experience of villagers, small farmers and others in Southern societies. Sir Albert Howard, often regarded as the originator of organic agriculture, derived many of his ideas from consulting with peasant cultivators in India whom he called his 'professors' (Howard 1953: 222), a strategy which was considered revolutionary in the context of British colonial administration. More recently, grassroots activists in Third World countries have emphasised the importance of 'ordinary knowledge' (Lind-blom and Cohen 1979) which depends more on keen observation and common sense than on professional techniques. This ordinary knowledge is accumulated within local grassroots networks by breathing air, drinking water, tilling soil, harvesting forest produce and fishing rivers, lakes and oceans (Breyman 1993: 131). In a similar fashion, native people in Northern societies accumulate firsthand knowledge of the environment which may not be available to non-indigenous observers. For example, it has been suggested[4] that biologists estimating the effect of mega projects on the ecology of rivers in the Canadian north may overlook the existence of a number of fish species simply because they never bother to ask native residents who know the land intimately (Richardson *et al.* 1993: 87).

In researching the origins of environmental claims, it is important for the researcher to ask where a claim comes from, who owns or manages it, what economic and political interests claims-makers represent and what type of resources they bring to the claims-making process.

In the early US conservation movement, environmental claims were largely traceable to an East Coast elite who utilised a network of 'old boy' ties to secure funding and political action. Enthusiastic amateurs, they dominated the boards of zoos, natural history museums and other public institutions from whence they were able to direct campaigns to save redwood trees, migratory birds, the American bison and other endangered species and habitats (Fox 1981). In a similar fashion, the threat to British birds, wildlife sites and other elements of nature was proclaimed in the late nineteenth and early twentieth centuries by a number of conservation groups with elite membership (Evans 1992; Sheail 1976).

By contrast, present day environmental claims-makers are more likely to take the form of professional social movements with paid administrative and research staffs, sophisticated fund-raising programmes and strong, institutionalised links both to legislators and the mass media. Some groups even use door-to-door canvassers who are paid an hourly wage or get to keep a percentage of their solicitations. Campaigns are planned in advance, often in pseudo-military fashion. Grassroots participation is not encouraged beyond 'paper memberships' with control centralised in the hands of a core group of full-time activists.

The process of assembling an environmental claim often involves a rough division of labour. While there are notable exceptions, research scientists are normally handicapped by a combination of scholarly caution, excessive use of technical jargon and inexperience in handling the media. As a result, an important finding may lie fallow for decades until proactively transformed into a claim by entrepreneurial organisations (Greenpeace, Friends of the Earth, Sierra Club) or individuals (Paul Ehrlich, Jeremy Rifkin). Greenpeace's claims-making activity, for example, does not so much flow out of its ability to construct entirely new environmental problems but rather from its genius in selecting, framing and elaborating scientific interpretations which might otherwise have gone unnoticed or been deliberately glossed over (Hansen 1993b: 171). Indeed, the nature of the relationship between the news media and environmental pressure groups such as Greenpeace has become sufficiently institutionalised (Anderson 1993: 55) that it would be difficult for an emergent problem to penetrate the mass media arena without at least token validation from the latter.

In assembling an environmental problem, not all explanations are created equally. Claims which hinge on difficult to understand concepts such as 'entropy' are far less likely to stick than those which have at their nucleus more readily comprehensible constructs; for example, 'extinction' or 'overpopulation'. Sometimes, the basic outline of a claim only becomes clear in the context of a political, economic or geographic 'crisis'. This was the case in 1973 when concerted action by OPEC (Organisation of Petroleum Exporting Countries), the oil producers' cartel, triggered an energy crisis in industrial nations in the West. Similarly, the abnormally hot US summer of 1988 gave the problem of global warming a visible, experiential focus.

Presenting environmental claims

In presenting an environmental claim, issue entrepreneurs have a dual mandate: they need both to command attention and to legitimise their claim (Solesbury 1976). While not unrelated, these constitute two quite separate tasks.

As Hilgartner and Bosk's (1988) model emphasises, the arenas through which social problems become defined and conveyed to the public are highly competitive. To command attention, a potential environmental problem must be seen to be novel, important and understandable – the same values which characterise news selection in general (Gans 1979).

One effective way of commanding attention is through the claimants' use of graphic, evocative verbal and visual imagery. Thus the extreme thinning of the ozone layer became much more saleable as an environmental problem when depicted as an expanding 'hole'; American children's entertainer, Bill Shontz, has even recorded a hit song entitled *Hole in the Ozone*. Similarly, the effects of acid rain were successfully dramatised when German environmentalists began to use the term *waldsterben* (forest die-back). Visual language can be especially powerful in carrying out this task. For example, technical data on the size of seal herds and codfish stocks instantly lost relevance when Brian Davies and other activists released photos to the media of baby seal pups being clubbed to death on the ice floes of Labrador.

It is not unusual, however, for these visual images to be streamlined so as to underline a central image. Mazur and Lee (1993: 711) give several striking examples of this. The NASA satellite pictures of the ozone hole over the Antarctic which became a 'logo' of the problem transformed continuous gradations in real ozone concentration into an ordinal scale that is colour-coded, conveying the erroneous impression that a discrete, identifiable hole could actually be located in the atmosphere over the South Pole. In August 1988, a *New York Times* article on rainforest destruction was accompanied by a stunning satellite photograph of the burning Amazon which was created by Alberto Setzer of the Brazilian Institute of Space Research. The photograph showed what appeared to be nearly 100,000 fires; however, it was really a composite of many separate pictures and included fires in areas of secondary forest growth as well as virgin rainforest.

Environmental issues may be forced into prominence when exemplified by particular incidents or events, for example, the nuclear accidents at Chernobyl and Three Mile Island, the Bhopal chemical disaster, the wreck of the oil-tankers *Torrey Canyon* and *Exxon Valdez*. Dramatic events like these are important because they assist political identification of the nature of an issue, the situations out of which it arises, the causes and effects, the identity of the activities and the groups in the community which are involved with the issue (Solesbury 1976: 384–5).

Staggenborg (1993) has identified six major types of 'critical events' which affect social movements such as the environmental movement. Large-scale socio-economic and political events such as wars, depressions and national elections influence the opportunities for collective action by altering perceptions of grievances and threats; for example, the 1980 election of US President Ronald Reagan led to increased memberships in environmental groups[5] since it raised the spectre of a free enterprise run rampant in national parks and other wilderness settings. National disasters and epidemics can represent a turning point in the movement, highlighting grievances and bringing about movement growth. Similarly, industrial and nuclear accidents can be potentially useful to the movement by laying bare policies and features of the power structure that are normally hidden; for example, the power of the oil companies in the Santa Barbara oil spill (Molotch 1970). Critical encounters involve face-to-face interaction between authorities and other movement actors focusing attention on movement issues. A recent example of this is the series of confrontations between protesters and police along the logging trails of Clayoquot Sound on Vancouver Island. Strategic initiatives are events created by deliberate actions taken by supporters or opponents to advance movement or counter-movement goals. The staged events which are characteristic of Greenpeace campaigns are examples of this, as is the publication of polemical books such as Paul Ehrlich's *The Population Bomb* and Jeremy Rifkin's *Beyond Beef*. Finally, policy outcomes are official responses to collective action by a movement or counter-movement – critical junctures at which movements are forced to renegotiate their strategies, tactics and goals as a result of changes in the political environment. The decision by the Roosevelt administration in 1914 to begin construction of the Hetch Hetchy Dam in Yosemite National Park in order to provide water for a pipeline to San Francisco was such a decision, in that it destroyed any possibility of a further alliance between the resource conservationists as represented by Gifford Pinchot and the preservationists led by John Muir.

Staggenborg's discussion is directed primarily towards the issues of social movement mobilisation and strategies, but her typology of events is relevant to the presentation of environmental claims insofar as environmental organisations often represent the primary claims-makers at this stage of the construction of environmental problems.

Of course, not all critical events are guaranteed to generate a high-profile problem. According to Enloe (1975: 21), an event provokes an environmental issue when it (1) stimulates media attention; (2) involves some arm of the government; (3) demands governmental decision; (4) is not written off by the public as a freak, one-time occurrence; and (5) relates to the personal interests of a significant number of citizens. These criteria are partly a function of the incident itself but also depend on the successful exploitation of the event by environmental promoters.

In presenting environmental claims, movement leaders engage in what Snow *et al.* (1986) have termed the process of 'frame alignment'; i.e. environmental groups tap into and manipulate existing public concerns and perceptions in order to broaden their appeal. For example, Greenpeace primarily chooses topics and organises campaigns in areas that can lend themselves to the widest public resonance (Eyerman and Jamison 1989: 112) while avoiding those which are more divisive. In a similar fashion, environmental movement opponents attempt to appeal to a wider public by linking new technologies or programmes to popular issues and causes. Thus the biotechnology industry has been successful in fostering a public image of an incremental and benign technology which is useful in promoting economic development (Plein 1991). Commanding attention is not, however, sufficient to get a new issue on to the agenda for public debate (Solesbury 1976: 387). Rather, emergent environmental problems must be legitimated in multiple arenas – the media, government, science and the public.

One way of achieving this legitimacy is through the use of the rhetorical tactics and strategies cited by Best (1987) and Ibarra and Kitsuse (1993). Rather than follow a chronological order, as Best suggests, environmental rhetoric has become increasingly polarised. Ecofeminists, deep ecologists and other critics of post-industrial society have tended to adopt a 'rhetoric of rectitude' which justifies consideration of environmental problems on strictly moral grounds. By contrast, environmental pragmatists, who advocate sundry versions of the 'sustainable development' paradigm, tend towards a rhetoric of rationality. Green business, for example, is based on the premise that environmentalism can be both socially useful and profitable.

This cleavage can be illustrated with reference to the loss of tropical rainforests in Brazil, Malaysia and Indonesia. Pragmatists argue that the loss of these rainforests is a serious problem because it leads to the extinction of rare indigenous insects, plants and animals which are invaluable to pharmaceutical companies as sources of new wonder drugs. Environmental purists, on the other hand, base their claims on a rhetoric which stresses the inherent spiritual value of these endangered habitats.[6]

Environmental claims can also be legitimated when their sponsors become legitimate and authoritative sources of information. Hansen (1993b) has demonstrated that Greenpeace has achieved this kind of sustained success as a claims-maker in a number of ways: by acting as a conduit for the dissemination of new scientific developments between the research community and the media; by becoming a 'shorthand signifier' for everything environmental – environmental caring, green lifestyles, environmentally conscious attitudes – and by producing knowledge and information which can be used strategically in public arena debates (see Eyerman and Jamison 1989).

It is sometimes possible to pinpoint an event which constitutes the turning point for an environmental problem and when it breaks through into the zone of legitimacy. With regard to global warming, this occurred at US Senate hearings in 1988 when Dr James Hansen made the claim that he was 99 per cent sure that the warming of the 1980s was not due to chance but rather to global warming. In the case of ozone depletion, the key event was a 1988 NASA/NOAA report providing hard evidence for the first time implicating CFCs (chlorofluorocarbons) in ozone layer depletion. With pulp mill dioxins, it was the 1987 release of the '5 Mill Study' showing that traces of this toxic chemical had been detected in various household paper products and the subsequent front page story in the *New York Times* which launched this problem in the United States, and, later, in Canada (Harrison and Hoberg 1991).

Yet scientific findings and testimony by themselves are not always sufficient to push an environmental problem past the break point of legitimacy. In the case of global warming, Dr Hansen's earlier Senate testimony in 1986, where he predicted that significant global warming might be felt within five to fifteen years, did not attract comparable coverage or concern. This only occurred two years later when there had been a significant shift in media practices and public attention (Ungar 1992: 492). Similarly, Molina and Rowland's 1974 publication in the journal *Nature* of their theory that CFCs were destroying the ozone layer at first only brought some limited coverage in the California press. It was only later on when the issue became linked to claims that other gases from aerosol cans, notably vinyl chloride, were linked to skin cancer, that their data were given wide attention and media legitimacy (Mazur and Lee 1993: 686).

Contesting environmental claims

Even if an emergent environmental claim manages to transcend the threshold of legitimacy, this does not automatically ensure that an amelio-rative action will be taken. As Gould *et al.* (1993: 229) have noted, one can interpret environmental protection history from the position that environmental movements have been far more succesful in getting listed on the broad political agenda than in getting their policies institutionalised within this agenda, especially where these policies might require the reallocation of resources away from large-scale capital interests and state bureaucratic actors.

Solesbury (1976: 392–5) has noted a number of factors which can contribute to an issue being lost at the point of decision or action. Major external constraints such as the onset of a national economic crisis may lead to a problem being postponed, then altogether abandoned. A problem may be transformed into a less threatening political issue. Opponents within government bureaucracies may use a number of tactics – postponing discussion, referring an item back for further research or amendment – which ensure that a problem will not immediately be acted upon.

As a consequence, invoking action on an environmental claim requires an ongoing contestation by claims-makers seeking to effect legal and political change. While scientific support and media attention continue to constitute an important part of the claim package, the problem is principally contested within the arena of politics. Contesting an environmental problem within the political policy stream is a fine art, given the cross pressures which legislators face.

Consider, for example, a recent article (Geddes 1994) on the shoals of conflicting interests which must be navigated by Canada's environment minister Sheila Copps as she prepares to introduce controversial new regulations on federal environmental assessments. If the regulators are too ambitious, Copps will 'run into a buzzsaw' from the provinces of Quebec and Alberta who will interpret this as an incursion of their constitutional powers. If the regulations are stringent, the business lobby, especially in the oil and gas sectors, will strenuously object. If she does too little, environmental groups will be critical, accusing her of failing to make good on a Liberal Party election promise to introduce much tougher regulations than those proposed by the former Conservative government.

Environmental entrepreneurs must skilfully guide their proposals through a log-jam of vested and often conflicting political interest groups, each of which is capable of stalling or sinking the proposals. As Walker has noted:

Public [environmental] policies seldom result from a rational process in which problems are precisely identified and then carefully matched with optimal solutions. Most policies emerge haltingly and piecemeal from a complicated series of bargains and compromises that reflect the biases, goals and enhancement needs of established agencies, professional communities and ambitious political entrepreneurs.

(1981:90)

Kingdon (1984) observes that policy proposals that survive in this political jungle usually satisfy several basic criteria.

First, legislators must be convinced that a proposal is technically feasible; that is, if enacted, the idea will work. This may not prove to be the case in hindsight; for example, the Endangered Species Act in the United States has worked out much less perfectly in its implementation than on paper. Nevertheless, a proposal must at least initially appear to be scientifically sound and politically administrable.

Second, a proposal which survives in the political community must be compatible with the values of policy-makers. Since most bureaucrats and politicians do not share the ecocentric views of American Vice-president Al Gore, this means that solutions which reflect the New Ecological Paradigm are not likely to get very far unless there is a generally perceived crisis. Instead, environmental solutions which on the surface appear to be neutral stand a better chance of being accepted than those which seem ideologically tinged. Furthermore, problems which are framed in utilitarian terms often go further than those which are not. This means that arguments made with financial expediency in mind – figures and statistics translated into 'bottom-line' dollars (pounds) – are more likely to resonate than those which are presented solely on the basis of moral justifications (Hunt *et al.* 1994: 200–1).

Environmental policy is by no means a perfectly predictable and consistent enterprise. For example, Milton (1991) has suggested that the British government routinely adopts a contradictory approach to the environment. On domestic pollution issues it adopts a rigid, hierarchical position which tends to retard change. This has been quite evident in, for example, the British response to the acid rain problem. By contrast, on international environmental problems such as global warming, the UK has adopted a more 'entrepreneurial' approach. On wildlife and conservation issues an approach which constitutes a mixture of the hierarchical and the entrepreneurial is favoured. Sometimes, an issue will rise in the policy agenda for totally unexpected reasons. This occurred with the greenhouse effect which initially achieved the stamp of seriousness not in terms of a long-range threat to the world climate but in relation to what was basically a side issue: the environmental implications of the large-scale deployment of the supersonic transport airplane (SST) in the early 1970s (Hart and Victor 1993: 663–4).

Thus successfully contesting an environmental claim in the political arena requires a unique blend of knowledge, timing and luck. This process is often event-driven with a disaster such as the Three Mile Island nuclear accident opening up 'political windows' (Kingdon 1984: 213) which would otherwise remain closed. This is not to say that agenda-setting and legislative action are totally random but the process is highly contingent upon a number of internal and external factors, many of which are not linked to the obvious merits of the case.

At the same time, there may also be a contest for 'ownership' of an environmental problem. This can be particularly rancorous where one of the contesting parties is drawn from the ranks of those directly victimised by a problem. There are many examples of this in the social problems field ranging from 'deviance liberation movements' such as the American prostitutes' rights campaign (Jenness 1993; Weitzer 1991) to victims' rights groups; for example, the one recently formed by breast cancer patients. This is less common with environmental problems which generally have a more diffused impact. One significant example, however, is the current dispute over the issue of who owns 'biodiversity' both as a resource and as an environmental problem (see Chapter 8). This struggle pits a coalition of small farmers, ecological activists and others in Third World countries against the conservation establishment: biologists, bureaucrats from non-governmental organisations and government ministries dealing with trade and environmental issues.

Hawkins (1993) has identified three ideal-type paradigms which occupy the increasingly contested discourse over global environmental futures. The prevailing 'global managerialist paradigm' advocates the detection and solution of problems in the globalised commons by an existing configuration of nation states and international organisations buttressed by scientific experts and professional environmentalists within international NGOs (non-governmental organisations). This approach downplays local perceptions and definitions of problems, and on occasion may even blame poor people in Third World nations for causing environmental degradation. The 'redistributive development paradigm' recognises the need for greater equity in matters pertaining to development and the environment in Southern countries. It proposes that such inequities can be redressed through a number of innovative measures such as the Green Fund within the World Bank or debt-for-nature swaps. The 'new international sustainability order paradigm' calls for a fundamental restructuring of the world order such that Third World nations claim a more direct voice in establishing a balance between economic and social sustainability.

Hawkins depicts the construction of international environmentalism as reflecting an ongoing struggle among supporters of these three paradigms. The dispute over the ownership of biodiversity is one recent manifestation of this; the conflict over global climate change is another. Even the language used in defining this contested ground is itself socially constructed. For example, countries of the North have adopted a 'globalised' language to describe the situation in Southern nations in which 'our' environmental problems (climate change, ozone depletion) are caused by 'their' develop-ment problems (forest loss, overpopulation), a situation which is solvable only by embracing 'sustainable development' strategies (Redclift and Woodgate 1994: 64–5). At present, the first two paradigms still predomi-nate but the new international sustainability order paradigm appears to be making some significant inroads.

Audiences for environmental claims

In addition to the skill of claims-makers and the severity of the underlying condition itself, the success of a putative environmental claim may also be tied to the magnitude of audiences that are mobilised around that claim. That is, a groundswell of audience support not only marks the rising awareness of a problem but it can also constitute a valuable resource in the effort to capture political attention.

For sociologists, the problem is how to reliably gauge the size and influence of audiences. As Ungar (1994: 298) has pointed out, the potential for environmental claims-makers to use public opinion as a resource is paradoxically both enhanced and limited by present polling procedures. That is, public polling today rarely maps support for contested positions, opting instead for broad measures of environmental concern such as the 'New Environmental Paradigm Scale' developed by Riley Dunlap and his colleagues. This produces such a vague barometer of public opinion that virtually any group on the 'pro-environmental' side can claim to represent it but, at the same time, it makes it difficult to gauge specific reactions to specific issues. Alternatively, one can look to other indicators of public support – recycling behaviour, green consumerism, participation in en-vironmental events and mobilisations – but these too are imperfect measures of opinion.

Nevertheless, a tide of public opinion can sweep a claim upwards on to the policy agenda, sometimes in a dramatic fashion. In the 'Alar' con-troversy in the United States, for example, public fears about toxins translated into a short-run consumer boycott of apples, even though the risk-supporting data were later found to be less reliable than was originally thought. Similarly, public concern about 'Mad Cow Disease' in Britain has been sufficiently grave for governments to have acted in a precautionary manner not always so evident in the case of other potential risks.

Of course, not all environmental claims succeed in raising the red flag for concerned audiences. Some claims are perceived as being too extreme, too misanthropic or too complex. Others run up against powerful counter-claims. Some fail because the requisite preventive or mitigative response mandates too great a lifestyle sacrifice.

In considering why some environmental claims capture the public eye and others do not, it may be helpful to look to the field of advertising research. In a recent large-scale comparative study which examined the attitudes of 30,000 consumers in twenty-one countries, the New York advertising agency Young & Rubicon came up with a marketing model, the 'Brand Asset Valuator', which isolates four key factors that predict how well a specific product will do in the market-place: uniqueness, relevance, stature and familiarity (Scotland 1994).

In the case of environmental claims, uniqueness or *distinctiveness* refers to the extent to which the public perceives a problem as separate from others of a similar nature. For example, acid rain claims-makers were successful in distinguishing this condition from the more inclusive category of air pollution. Rhetorical strategies are important here in creating distinctive labels for emerging problems as well as devising symbolic codes which can be attached to a claim in order to confer a distinctive identity.

Relevance refers to the degree to which a particular environmental problem matters to the ordinary citizen. This is not always easy to demonstrate, even when the problem is occurring in people's own back-yards. It is especially difficult in the case of global environmental problems which have their origins far away in distant parts of the world. Thus extended drought conditions in the poor African nations are of little relevance in the South-western United States, yet regional water shortages which require that local citizens stop watering their lawns and filling their swimming pools are quite meaningful.

Stature denotes how highly a consumer thinks and feels about a particular brand. In the case of the environment, this refers to the attitudes of the public towards the place or people or species under threat. It is no accident that the wildlife protection movement first mobilised in the nineteenth century over the danger posed to our much-loved songbirds by hunters and by the millinery trade. Similarly, national parks and monuments – Yellowstone Park in the United States, the Lake District in Britain, the

Great Barrier Reef in Australia – have considerable symbolic stature which comes into play if these places are imperilled. By contrast, low income black and Hispanic communities in the American South which face serious threats from toxic polluters have long been accorded low stature, especially by middle-class audiences.

Finally, *familiarity* refers to how well known a particular problem is to an audience. The media play an especially important role here in educating us about environments, species or places which may have been beyond our realm of personal experience. For example, in 1992 it was announced that scientists in Central Vietnam had discovered the *sao la*, a goat-like mammal previously unknown to the outside world. Almost overnight, the *sao la* became a media superstar as a result of a media frenzy whipped up by scientists, environmentalists and the press.[7] Celebrated on the pages of the *National Geographic* and *People* magazines, it became 'the zoological equivalent of finding a new planet' (Shenon 1994). In some cases, environmental activists may undertake collective action in order to familiarise audiences with a claim. For example, the clear-cutting practices in the old growth forests in British Columbia have recently become widely known in Europe and America, in part because of the extensive media coverage of protests by environmental activists on the logging roads and on the steps of the provincial legislature. Rather than enhancing the stature of a claim, however, familiarity may ultimately produce issue fatigue on the part of the general public, especially if new developments are not forthcoming. This is the case even if a problem is both distinctive and relevant. Indeed, audiences have an inherent sense of fair play which dictates that activities such as unrelenting 'polluter bashing' are unacceptable, even if the criticism is well deserved.

Successful environmental claims, then, must possess elements of vitality and stature which ensure that they will not perish in a sea of disinterest or irrelevance.

Necessary factors for the successful construction of an environmental problem

It is possible to identify six factors which are necessary for the successful construction of an environmental problem. These are as follows.

First, an environmental problem must have scientific authority for and validation of its claims. Science may well be an 'unreliable friend' to the environmental movement as Yearley (1992) has suggested, but nevertheless it is virtually impossible for an environmental condition to be successfully transformed into a problem without a confirming body of data which

1   Scientific authority for and validation of claims.
2   Existence of 'popularisers' who can bridge environmentalism and science.
3   Media attention in which the problem is 'framed' as novel and important.
4   Dramatisation of the problem in symbolic and visual terms.
5   Economic incentives for taking positive action.
6   Emergence of an institutional sponsor who can ensure both legitimacy and continuity.

comes from the physical or life sciences. This is especially so with the newer global environmental problems, whose very existence hinges on a novel scientific construction (see the discussion of biodiversity loss in Chapter 8).

Second, it is crucial to have one or more scientific 'popularisers' who can transform what would otherwise remain a fascinating but esoteric piece of research into a proactive environmental claim. In some cases (e.g. Edward Wilson, Paul Ehrlich, Barry Commoner), the popularisers may themselves be employed as scientists; in others (e.g. Jonathan Porritt, Jeremy Rifkin), they are activist authors whose knowledge of science comes secondhand. Whatever their background, these popularisers assume the role of entrepreneurs, reframing and packaging claims so that they appeal to editors, journalists, political leaders and other opinion-makers.

Third, a prospective environmental problem must receive media attention in which the relevant claim is 'framed' as both real and important. This has been the case for most of the well-known problems of the present decade; for example, ozone depletion, biodiversity loss, rainforest destruction, global warming. By contrast, other significant environmental problems fail to make the public agenda because they are not considered especially newsworthy. For example, in many Canadian cities lack of treatment of urban sewage is endemic but this has received scant coverage compared to other pollution problems. As the executive director of the Sierra Legal Defense Fund recently pointed out, a volume of sewage equivalent to thirty-two oil-tankers the size of the *Exxon Valdez* is dumped each day into local rivers or bays, yet it is done out of the sight of the public with virtually no attention from the media (Westell 1994).

Fourth, a potential environmental problem must be dramatised in highly symbolic and visual terms. Ozone depletion was not a candidate for widespread public concern until the decline in concentration was graphically depicted as a hole over the Antarctic. The wanton practices of the major forestry companies only became a matter for international outrage

55

when Greenpeace and other environmental groups began to exhibit dramatic photographs of the 'clear-cuts' on Vancouver Island while labelling the area the 'Brazil of the North'. Images such as this provide a kind of cognitive short cut compressing a complex argument into one which is easily comprehensible and ethically stimulating.

Fifth, there must be visible economic incentives for taking action on an environmental problem. This is so in each of the three detailed case histories which will be presented in future chapters in this volume. In the case of acid rain, a variety of economic interest groups, from the forest farmers and hunting associations of the Black Forest in Southern Germany to the maple syrup producers of New England and Eastern Canada, supported the claim by scientists and environmentalists that sulphur dioxide emissions from smelters and power plants were causing the forest to die back. The case for acting boldly to stop biodiversity loss was levered on the argument that the tropical rainforests contained an untapped wealth of pharmaceuticals which would disappear for ever if nothing was done. The campaign against the genetically engineered hormone bST has a backbone of strong support from farmers and legislators in several Northern states in which dairy producers face a loss of income if this biotechnology becomes widely utilised. At the same time, environmental claims which carry positive, economic incentives for one group may also involve costs for others, thus provoking sharp opposition. This was certainly the case with acid rain; sulphur dioxide producers such as the Central Electricity Generating Board in Britain vigorously contested the definition and parameters of the problem as presented by environmental claims-makers.

Finally, for a prospective environmental problem to be fully and successfully contested, there should be an institutional sponsor who can ensure both legitimacy and continuity. This is especially important once a problem has made the policy agenda and legislation is sought. Internationally, this can be seen in the important role played by agencies and NGOs associated with the United Nations.

CONCLUSION

I will elaborate this social constructionist view of the environment in Chapters 3–6, starting with an analysis of the emergent and collaborative nature of both environmental risks and knowledge. I contend that the concept of environmentalism itself is a multi-faceted construction which welds together a clutch of philosophies, ideologies, scientific specialties and policy initiatives. As part of this discussion, separate chapters are devoted both to the central role of media discourse and that of science in interpret-

ing and shaping the contexts, conditions and consequences of the environmental crisis.

I then present three case studies to illustrate the dynamics through which
environmental problems are divined, defined, legitimated and contested.
The first of these, acid rain, is a mature environmental problem which has
been in the public eye for more than a quarter of a century. Unlike global
warming or ozone depletion, its effects have been more continental than
global, although evidence of acid precipitation has been found as far away
as Antarctica. A second, more recent problem, 'biodiversity loss', has
become one of the high-profile global environmental problems of the
1990s as evidenced by its featured status at the June 1992 Rio Conference.
It does not physically impact the entire planet directly at one and the same
time as do other global environmental problems. Nevertheless, the politics
and economics of biodiversity loss are such that they span North and South,
thus requiring solutions within the international political arena. A third
problem, biotechnology as an environmental risk, is examined through the
specific case of recombinant bovine somatotropin (bST), a bioengineered
hormone which has recently been approved for use in the American dairy
industry. The bST[8] controversy offers an example where a furious contest
continues to rage over whether the concerns of environmentalists, animal
rights activists and other claims-makers will be judged valid or not.
Although it has thus far been restricted primarily to a national setting, the
highly interrelated nature of the global agri-food system implies that
pressures for or against its use in agriculture will inevitably transcend
national borders.

In the final chapter, the social construction of environmental risk and
knowledge is examined in the context of the wider theoretical debate
within sociology over the trajectory of contemporary societies towards a
'late modern' versus a 'postmodern' future. After analysing two major
attempts to conceptualise ecological concerns within a broader societal
context – Ulrich Beck's theory of the 'risk society' and Mol and Spaargaren's version of 'ecological modernisation theory' – I suggest several
ways in which environmental constructionism and the postmodern condition interweave. The book concludes with some suggestions for the
future course of research on the environment from a social constructionist
perspective.

3

NEWS MEDIA AND ENVIRONMENTAL COMMUNICATION

In moving environmental problems from conditions to issues to policy concerns, media visibility is crucial. Without media coverage it is unlikely that an erstwhile problem will either enter into the arena of public discourse or become part of the political process. In fact, most of us depend on the media to help make sense of the often bewildering daily deluge of information about environmental risks, technologies and initiatives.[1]

Yet at the same time, the media's role as an agent of environmental education and agenda-setting is quite complex. As Schoenfeld *et al.* (1979) have demonstrated, the daily press in the United States was initially slow in grasping the basic substance and style of environmentalism, leaving it to issue entrepreneurs in colleges and universities, government and public interest groups to mobilise concern outside of the media net. In local environmental conflicts, media claims are often viewed sceptically, re-fracted as they are through the prism of residents' own practical everyday experiences and knowledge (Burgess and Harrison 1993). Rather than actively sparking a response to environmental problems, the media often seem to be a millstone weighing down public discussion of environmental topics in a technical–bureaucratic discourse which excludes interest groups and non-official claims-makers (Corbett 1993: 82).

In this chapter, I will assess the news media's[2] conflicting role in socially constructing environmental issues and problems. Of particular concern is the extent to which the portrait of the environment presented by main-stream journalists represents a critique of the paradigm of technological progress as opposed to simply an extension of the existing corpus of disaster stories. First, however, it is necessary to briefly outline the general process through which the media 'manufacture' news and endow issues and events with symbolic meaning.

MANUFACTURING NEWS

For many years, mass communication researchers largely took for granted the existence of 'objective' facts and events which could be verified, exclusive of whether or not they were actually covered by the media. Thus floods and hurricanes, political victories and resignations, medical miracles and foreign wars were all thought to have a certifiable existence of their own beyond the newsroom. Reporters, editors, producers and other 'newsworkers' might on occasion distort or selectively omit certain happenings but this did not mean that they were not real (Fishman 1980: 13).

In the 1970s, this approach gave way to a very different model, in which events become news only when transformed by the newswork process and not because of their objective characteristics (Altheide 1976: 173). News is conceptualised here as a 'constructed reality' in which journalists define and redefine social meanings as part of their everyday working routine (Tuchman 1978). News-making, in turn, is treated as a collaborative process in which journalists and their sources negotiate stories.

Organisational routines and constraints

While the construction of news may be influenced by cultural or political factors, it is generally seen as the result of inescapable organisational routines within the newsroom itself. Schlesinger (1978) observes that rather than being a form of 'recurring accident', news is the product of a fixed system of work whose goal is to impose a sense of order and predictability upon the chaos of multiple, often unrelated events and issues. In his observational study of BBC news, he found that the backbone of each day's newscasts was a 'routine agenda of predictable stories': labour negotiations, parliamentary business, activities of the Royal Family, sport scores, etc. In a similar fashion, Fishman (1980) observed that, rather than dig for information, reporters at a California daily newspaper opted instead for a diet of routine news derived from a mix of scheduled events (press conferences, courtroom trials) and pre-formulated accounts of events (arrest records, press releases); these items were crucial in helping them to meet deadlines and story quotas.

In addition to mandating that news be planned, time also acts as a constraint upon the final product itself. This has the effect of rendering news reports 'incomprehensible rather than comprehensive' (Clarke 1981: 43). In particular, action clips which fit more easily into existing formats, especially television news, are favoured over longer, more nuanced stories which deal with underlying causes and conditions.

Furthermore, by consistently failing to ask the question 'why' the news process 'contributes to *decontextualising*, or removing an event from the context in which it occurs in order to *recontextualise* within news formats' (Altheide 1976: 179). This tendency is further encouraged by the use of news 'angles' – frameworks around which a particular content is moulded in order to tell a story. The use of news angles is pervasive in journalism and plays a significant role in determining not only the 'spin' put on a story but also whether a story is suitable in the first place for broadcast or publication.

Media constructionists have also noted the importance of news sources in shaping story content. Reporters usually stick to a short-list of trusted source contacts who, on the basis of past experience, can be counted on to be both articulate and reliable. In fact, it is not unknown for source contact lists to be passed down from one reporter to the next. Trusted sources come from various walks of life but they are usually people who function in official roles: politicians, the heads of government agencies, scientists and other experts. Even where the media solicit comment from opponents of the *status quo*, news sources are invariably drawn from the executive of major social movement organisations such as Greenpeace and Friends of the Earth.

In their study of the 1969 oil spill in Santa Barbara, California, Molotch and Lester (1975) found that powerful figures and organisations with routine access to the media (the President, federal officials, oil company representatives) were far more likely to function as news sources than were conservationists and local officials. These sources exercise considerable social and political power by providing a pre-packaged, self-serving, socially constructed interpretation of a given set of events or circumstances – an interpretation that is readily adopted by journalists who rarely have the time or the specialised knowledge needed to flesh out their own news angle (Smith 1992: 28).

Media discourse

In recent years, media constructionists have looked beyond the social organisation of the newsroom and focused on the process by which journalists and other cultural entrepreneurs develop and crystallise meaning in public discourse (Gamson and Modigliani 1989: 2). This approach takes as its central concern the decoding of media texts – the visual imagery, sound and language produced in the social construction of news and other forms of public communication (Gamson et al. 1992: 381).

The key element here is that of *media frames*, a concept adapted by several media sociologists in the late 1970s and early 1980s (Gitlin 1980; Tuchman 1978) from Erving Goffman's work on small group interaction. Frames, like news angles, are organising devices that help both the journalist and the public make sense of issues and events and thereby inject them with meaning. In short, they furnish an answer to the question 'What is it that's going on here?' (Benford 1993: 678). When expressed over a duration of time, frames are known as 'story lines' (Gamson and Wolfsfeld 1993: 118).

Even when the details of an event are not disputed, the event can be framed in a number of different ways. For example, the 1993 murder of Liverpool toddler James Bulger by two 10-year-old boys was variously framed by the press as a new low in the continuing economic and moral decline of England, the turning point in the campaign against 'video nasties' (one boy's father had reportedly rented the movie *Child's Play 3* just before the crime), a cautionary tale for harried parents with youngsters in tow and an example of the linkage between school truancy and juvenile crime. Both claims-makers and their opponents routinely compete to promote their favoured frames to journalists as well as to potential supporters. At the same time, newsworkers forge their own frames largely for reasons of efficiency and story suitability. Gamson and Wolfsfeld (1993) depict the interaction between movements and the media as a subtle 'contest over meaning' in which activists attempt to 'sell' their preferred images, arguments and story lines to journalists and editors who, more often than not, prefer to maintain and reproduce the dominant mainstream frames and cultural codes. In the Nicaraguan conflict of the 1980s, for example, peace activists attempted to counter the official frame that the American-sponsored Contras were waging a struggle against Communist expansion by promoting a 'human costs of war are too high frame' (Ryan 1991).

Finally, as Gamson et al. (1992: 384) point out, it is wrong to assume that news consumers (readers, audiences) passively accept media frames as they are; they too may decode media images in different ways utilising varying frameworks of interpretation (Corner and Richardson 1993).

Media discourse, therefore, takes the form of a 'symbolic contest' in which competing sponsors of different frames measure their success by gauging how well their preferred meanings and interpretations are doing in various media arenas (Gamson et al. 1992: 385).

The process of framing is in many ways comparable to the rhetoric of claims-making in social problem construction (see Chapter 2). Gamson and Modigliani (1989: 3–4) distinguish five *framing devices*: metaphors, exemplars (i.e. historical examples from which lessons are drawn), catch-phrases, depictions and visual images, and three *reasoning devices*: roots (a causal analysis), consequences (i.e. a particular type of effect) and appeals to principle (a set of moral claims) which function as a kind of symbolic shorthand in telegraphing the core meaning of a frame.

Furthermore, they introduce the concept of *media packages*. Media packages help to organise these framing devices in cases of complex policy issues such as the use of nuclear power. In analysing television news coverage, news magazine accounts, editorial cartoons and syndicated opinion columns on nuclear power from 1945 to the present day, Gamson and Modigliani isolate seven different interpretive packages: progress; energy independence; the devil's bargain; runaway; public accountability; not cost-effective; and soft paths. As the titles suggest, each package is represented by a 'deft metaphor, catchphrase or other symbolic device' (1989: 3).

MASS MEDIA AND ENVIRONMENTAL NEWS COVERAGE

As Schoenfeld *et al.* (1979: 42–3) have demonstrated, prior to 1969, the daily press in the United States had considerable difficulty recognising environmentalism as a topic separate from that of conservation. Conservation was a reasonably well-understood and respectable concern, having been around since the 1880s. It had a known constituency, its own legislative acts and administrative bureaux and even its own universally recognized symbol – 'Smokey the Bear'. By contrast, the central tenet of environmentalism, i.e. that 'everything is connected to everything else', seemed difficult to grasp in journalistic terms. Similarly, in Britain, the preservation of the countryside, the national heritage and rare species of fauna and flora were all widely accepted as legitimate activities which cut across class lines, but few journalists readily connected them with air pollution, oil spills and other contemporary environmental problems.

During the late 1960s and early 1970s, media coverage of the environment rose dramatically and, for the first time, environmental issues were seen by journalists in both Britain and America as a major category of news (Lacey and Longman 1993; Parlour and Schatzow 1978). Newsworkers began to perceive individual difficulties such as traffic problems or pollution incidents as part of a more general problem of 'the environment' (Brookes *et al.* 1976; Lowe and Morrison 1984).

There are several key events which may be cited in order to explain this upswing in media awareness and understanding of environmentalists' claims. Schoenfeld *et al.* (1979: 43), citing Roth (1978), argue that the single most effective environmental message of the century was totally inadvertent: the 1969 view from the moon of a fragile, finite 'Spaceship

Earth'.[3] This provided a powerful metaphor with which to frame the environmental message. Similarly, Earth Day 1970 acted as a news 'peg' for a variety of otherwise disparate news stories on environmentally related subjects earning extensive coverage both nationally and in many local American communities (Morrison *et al.* 1972).

After 1970, however, media coverage of the environment began to fall off (Parlour and Schatzow 1978), although it recovered briefly during the energy crisis of 1973–4. When environmental stories did appear they were most likely to be event-related and problem-specific. In their examination of article headlines in the *Canadian Newspaper Index*, Einsiedal and Coughlan (1993: 140) observed that environmental items were located under a series of disparate and seemingly unconnected problem categories: air pollution; water pollution; waste management; and wildlife conservation. Similarly, Hansen (1993a: xvi) notes the tendency of the media to define the environment 'largely in terms of anything nuclear (nuclear power, nuclear radiation, nuclear waste, nuclear weapons), in terms of pollution and in terms of conservation/protection of endangered species'. Rarely were the global aspects of environmental problems highlighted during this period. Even more unusual was the appearance of stories on environmental problems in Third World countries.

This pattern appears to have changed somewhat over the last decade. Einsiedal and Coughlan note that towards the end of 1983 new descriptor terms began to appear in Canadian newspaper headlines; for example, 'global catastrophe', 'environmental order' and 'environmental ethics'. In contrast to the earlier conservation focus, environmental stories were vested with a more global character, encompassing attributes that included 'holism and interdependence and the finiteness of resources' (1993: 141). They also note an increasing urgency and seriousness in the coverage of environmental issues by the Canadian press as indicated by the appearance of a collection of 'war and dominance' metaphors: survival, defeat, battles, crusades. Topic headings were found to be more specific, covering such areas as 'eco-tourism', 'environmental law' and 'eco-feminism'. In a similar fashion, Howenstine (1987) notes a transformation in environmental reporting in major US periodicals from 1970 to 1982 towards a greater complexity of coverage. In addition, he found a shift across time to relatively fewer articles on the degradation and protection of the natural environment and more on economic and developmental issues.[4]

However, this perception that environmental coverage is deepening may be overly optimistic. Lacey and Longman (1993) note the rise of a 'show business' and commercial approach to environmental issues in the British media during the 1980s and argue that the improvements in environmental reporting are only evident if a narrow definition of environmental issues is utilised. In particular, an artificial separation is created between the environment and development issues in line with a predominant editorial and political bias. For example, recent coverage of famine in East Africa has been long on shock tactics but short on political insight, especially in the case of the drought in the Sudan, a country whose political regime is considered ideologically unacceptable by Western policy-brokers.

Production of environmental news

To a large extent, media coverage of environmental issues is constrained and shaped by the same production constraints which govern newswork in general. Earlier in this chapter, we discussed some of the most significant of these: limited production periods; limited story lengths; and limited sources. Clarke (1992) has grouped these production constraints into two general categories: short-term logistical and technological constraints, and long-term economic and occupational constraints which are embedded in the news process itself.

Short-term pressures of time have meant that environmental issues and problems have often been framed by journalists within an event orientation. As Dunwoody and Griffin (1993: 47) point out, this event orientation limits journalistic frames in two ways: (1) it allows news sources to control the establishment of story frames; (2) it absolves journalists from attending to the bigger environmental picture. Three major types of environmental events can be identified: milestones (Earth Day, the Rio Summit); catastrophes (oil spills, nuclear accidents, toxic fires); and legal/administrative happenings (parliamentary hearings, trials, release of environmental white papers).

The twin lures of celebrity and symbolism at milestone events can be seen at the 1992 Earth Summit at Rio de Janeiro in Brazil. Those attending included not only more than a hundred heads of state, including US President George Bush, British Prime Minister John Major, German Chancellor Helmut Kohl and Cuban President Fidel Castro, but also an estimated 12,000 representatives from NGOs. Among the celebrities from the world of politics and entertainment were former California governor Jerry Brown, actors Jeremy Irons and Jane Fonda and American media mogul Ted Turner. Even before the official summit began, a fundamental conflict arose between the wealthy nations of the North and the poorer countries of the South over a wide spectrum of issues. Finally, the summit was accompanied by an array of what *Time* magazine called 'sideshows galore' (Dorfman 1992): a fantasy ballet, *Forest of the Amazon*, an indigenous people's conference and a concert for the Life of Planet Earth. The symbolism of the occasion was typified by the giant Tree of Life in Rio on which were hung leaf postcards from children worldwide.

Environmental catastrophes are the bread and butter of environmental news coverage. They frequently involve injury and loss of life or the possibility of such. There are sometimes acts of tremendous courage or self-sacrifice. Human interest stories abound: the stubborn but proud homeowner who sits on the roof and refuses to evacuate as the floodwaters rise; the baby who is found alive after three days in the rubble of an earthquake-devastated neighbourhood.

According to Wilkins and Patterson (1990: 19), this event-centred reporting is characteristic not only of quick onset disasters such as tornadoes, hurricanes and blizzards but also of slow onset environmental hazards: global warming, ozone depletion, acid rain, and so on. In order to fit these latter phenomena into the news agenda, journalists are required to picture them as the recent outcome of an event rather than the inevitable outcome of a series of political and societal decisions.

While event-centred coverage has the advantage of raising public awareness of otherwise ignored environmental topics, it also has a negative side. By focusing on discrete events rather than on the contexts in which they occur, the media tend to give news consumers the impression that individuals or errant corporations rather than institutional politics and social developments are responsible for those events (Smith 1992; Wilkins and Patterson 1990). This is especially the case with environmental catastrophes. For example, in the case of the 1989 *Exxon Valdez* oil spill, the media framed the story in terms of Captain Joseph Hazelwood's alleged alcohol problems rather than deal with other potentially important news angles such as the recent history of cutbacks in maritime safety standards administered by the coast guard, or the oil industry's lack of capability in cleaning up large oil spills in settings such as Prince William Sound (Smith 1992). Cottle (1993: 122) has described this as the tendency of an item to remain 'entrapped within the narrow confines of its news format', unable to allow any background explanation or any input from outside, non-official voices.

Furthermore, media stories on environmental hazard events favour monocausal frames rather than frames involving long and complex causal networks. Thus Spencer and Triche (1994) found that increases in toxic pollution in the drinking-water supply of New Orleans during the summer of 1988 were almost exclusively attributed to a simple natural phenomenon – a drop in water levels in the Mississippi River due to drought conditions

– rather than to a combination of low water levels and a long-standing problem with discharges from chemical plants upriver from the city. They speculate that this monocausal framing occurred because newspaper personnel were reluctant to implicate several powerful institutional actors – the US Army Corps of Engineers, the state bureaucracy, the chemical industry – as contributors to this hazard event.

Cottle's comments further suggest a second feature of the news process which shapes the nature of environmental coverage: a public access which is largely restricted to official news sources. Since few reporters themselves feel qualified to sort out the often conflicting scientific, technical and political claims involved in an environmental problem, they either avoid substantive issues altogether (Nelkin 1987) or turn to informed sources[5] who can offer a credible and easily summarised précis of what is happening.

While these 'primary definers' are often depicted as coming exclusively from a hierarchy of social and political elites, Cottle (1993: 12) argues that this is not necessarily the case for environmental stories.. Analysing a sample of British television programmes from 1991 to 1992, he found that various diverse elements (i.e. scientists, diplomats, local officials and politicians, environmental pressure groups, individual citizens) collectively constituted the primary definers.[6] At the same time, Cottle indicates that this was by no means 'a situation of open and equal access' since environmental news clearly depends on a number of well-organised interests, some from the dominant elite, some from opposing groups.

However, Anderson (1993) has questionned whether it is possible to deduce patterns of source dependence from content analysis alone. Supplementing content analysis with interviews, she found that ease of access varies over time. For example, during the late 1980s, Greenpeace and Friends of the Earth had good access to the national media in Britain, but they have experienced some difficulties since because the environment has been overshadowed by other issues such as the economic recession.

At various points in its recent history, environmental news coverage has also suffered because it does not fit easily into the structure of routine news production. Metropolitan daily newspapers tend to be partially organised according to fixed 'beats' – city hall, industrial (labour) relations, crime, sports, etc. Schoenfeld (1980: 458) cites one reporter as describing the classic environmental story as a 'business–medical–scientific–economic–political–social–pollution story'. This being so, editors and producers often do not know what to do with stories about the environment. It should be noted, however, that this can have a positive aspect, insomuch as individual environmental reporters are sometimes given considerably more leeway than their colleagues working on other journalistic beats because environmental issues are so often difficult for non-specialists to understand (Fletcher and Stahlbrand 1992: 183).

Smaller newspapers and broadcast newsrooms are less likely to use beats, opting instead for a general assignment system (Friedman 1984: 4). This, however, creates its own difficulties. General assignment reporters, despite their optimism that they can quickly acquire adequate knowledge about subjects in which they have no background or training, are rarely capable of sophisticated reporting[7] such as that demanded by many environmental stories.

Based on his comprehensive analysis of news coverage of three environmental catastrophes – the 1988 Yellowstone Park forest fires, the *Exxon Valdez* oil spill and the Loma Prieta 'World Series' earthquake in 1989 – Conrad Smith, himself a former television photographer and film editor, identifies three major difficulties experienced by such general reporters: (1) they did not conceptualise these major catastrophes as anything more than large-scale versions of warehouse fires or train derailments; (2) they did not have the structural freedom to go beyond the obvious stories; (3) they did not know how to find experts and evaluate their relative scientific qualifications (1992: 190).

When environmentalism took off as a news story in 1969–70, many daily newspapers set up an ecology or environmental beat.[8] Reporters were recruited from allied beats – nature, outdoor recreation, science – or from the general assignment pool. While the volume of environmental coverage rose, the quality did not always keep pace. In particular, these new environmental reporters seemed to experience difficulty with both the substance and style of environmentalism (Schoenfeld *et al.* 1979). When the environment faded as an issue after 1970 many of these beats were shut down (Friedman 1983), although some of them have recently been recommissioned (Hansen 1991).

A final short-term constraint on environmental reporting is the role and influence of news editors. With one eye always fixed on circulation or audience figures, editors favour stories which feature controversy and conflict. As a result, thoughtfulness often gives way to sensationalism. In addition, editors are more likely to be sensitive to external pressures from corporate advertisers and other powerful supporters of the *status quo*. Reporters know this, and on occasion may modify or deliberately overlook significant stories which involve environmental wrongdoing (Friedman 1983). This occurred in the late 1970s in Houston, Texas, where local newspaper reporters were not willing to go against the predominant 'boomtown' mentality and report the problems surrounding a nuclear power plant and a nuclear waste treatment facility (Hochberg 1980).

Longer-term constraints on environmental journalism relate to historically evolved journalistic priorities, notably the requirements for news 'balance' and 'objectivity'. These dual pillars of objective journalism first arose during the nineteenth century as part of the sweeping intellectual movement towards scientific detachment and the culture-wide separation of fact from value (Gitlin 1980: 268). Despite periodic lapses, newsworkers today view objectivity and balance as the cornerstones of their profession.[9]

For environmental reporting, objectivity and balance mean that reporters often attempt to distance themselves and their readers from the environmentalist struggle to effect a shift in public consciousness, taking refuge instead in the objectivism of science (Killingsworth and Palmer 1992: 149). Journalists thus see themselves as a neutral and ironic voice, willing to be won over only if the scientific evidence concerning acid rain, global warming, biotechnology, etc., is sufficiently powerful and unambiguous.

The major shortcoming of this approach is that few environmental reporters are sufficiently well informed to be able to effectively evaluate the 'scientific standing' (Friedman 1983: 25) of the evidence. Alternatively, reporters may turn to the traditional 'equal time' technique whereby both environmentalist claims-makers and their opponents are quoted with no attempt to resolve who is right. In this case it becomes difficult for environmentalists to convince the public that an 'issue' is in fact a 'problem'.

The ideal of objectivity also means that journalists rarely express the content of environmental stories in overtly political terms, opting instead for news frames which emphasise conservation, civic responsibility and consumerism. Lowe and Morrison (1984: 80) even go so far as to contend that a major attraction of environmental issues for the media is that they can be depicted in non-partisan terms, allowing journalists to subversively foster environmental protest at the same time as appearing to maintain a politically balanced stance.

Cottle (1993: 128) echoes Habermas in noting how the media debase the public sphere, refracting the environment through a journalistic prism which reduces politically charged stories such as global warming to the more immediate and mundane domestic and leisure concerns of ordinary consumers; for example, whether a beach holiday is likely this summer.

Constructing 'winning' environmental accounts

As Stallings (1990: 88) has noted, some media accounts of environmental problems drop by the wayside while other 'winning accounts' persist and ultimately succeed in gaining acceptance. Indeed, the media contribute to this by fostering an image of either growth or decay for a particular problem (Downs 1972). In charting the ascent and tenure of environmental problems on the media agenda, it is possible to identify five key factors.

First, in order to gain prominence, a potential problem must be cast in terms which 'resonate' with existing and widely held cultural concepts (Kunst and Witlox 1993: 4). This is why the frame alignment process discussed in Chapter 2 is so crucial. Despite a quarter-century of exposure to environmental discourse, the actual awareness and salience of most environmental issues remains 'pitifully low' (Cantrill 1992: 37). In particular, most citizens continue to place their faith in science and technology and to believe that economic growth is generally desirable. Thus, packaging an issue in the form of direct criticism of the Dominant Social Paradigm would not appear to be an effective communication strategy for environmental claims-makers. Instead, it makes more sense to situate environmental messages in frames which have wider recognition and support in the target population: health and safety, bureaucratic bungling, good citizenship, and so on.

Second, a potential environmental problem must be articulated through the agendas of established 'authority fora' (Hansen 1991:451), notably politics and science. If it does not receive this legitimation, a problem will likely stagnate outside the media arena. This was the case in Britain where various 'green' issues (acid rain, ozone damage) lay relatively fallow until invigorated by a speech by Prime Minister Margaret Thatcher to the Royal Society in September 1988, in which she adopted an environmental rhetoric for the first time. The Thatcher speech conferred a new degree of political legitimacy on the environment and the environmental movement and this subsequently diffused throughout many other arenas with the assistance of the mass media (Cracknell 1993).

Third, environmental problems which conform to the model of a publicly staged 'social drama' are more likely to engage the attention of the media than those which do not. As Palmlund (1992:199) has suggested, the societal evaluation of risk takes the form of a dramatic contest coloured by emotions and containing both blaming games and games of celebration. In this contest there are readily identifiable heroes, villains, victims and even a chorus. Love Canal was the perfect media story by this yardstick with the timid housewife turned activist Lois Gibbs as the heroine, neighbourhood children with their increasing health problems as the primary victims, and Hooker Chemical as the odious polluter.

Some environmental organisations, notably Greenpeace, have been very successful in staging morality plays in front of the global media with themselves as intrepid idealists and a changing cast of characters – whalers, seal hunters, French sailors, nuclear operators – as the villains. By contrast, problems which lack this fairy-tale quality, for example, the seepage of indoor radon gas into Canadian and American homes, are more difficult although not impossible to sell to the media.

Fourth, an environmental problem must be able to be related to the present rather than the distant future in order to capture media attention. Dianne Dumanoski, environmental reporter for the *Boston Globe*, notes that some of the more immediate environmental problems such as oil spills interest editors more 'because they can understand that. . . . There's dirty stuff on the rocks; it's not computer models and these guys at MIT talking about something in the future' (Stocking and Leonard 1990: 41).

Global warming appeared to be a far away problem until the abnormally hot summer of 1988 when a series of tangible environmental disasters – droughts, floods, forest fires, polluted beaches – dominated the news. These contributed significantly to *Time* magazine's editorial decision to feature the endangered earth in its Planet of the Year issue of 2 January 1989 (McManus 1989).

Finally, an environmental problem should have an 'action agenda' attached to it either at the international (global conventions, treaties, programmes) or the local community (tree-planting, recycling) level. Environmental conditions which are less amenable to action are not as likely to appeal to reporters and editors unless, as was the case with the Ethiopian famine, a moral panic can be created around the consequences provoking a flurry of humanitarian relief efforts. Furthermore, rather than advocating some long-term action plan with results which may not be noticed for decades, environmental claims-makers should be able to offer the media some tangible results in the here and now: for example, shutting down an incinerator, cleaning up a polluted harbour, rescuing a beached whale. Unfortunately, as Solesbury (1976: 395) has noted, complex environmental problems with multiple dimensions are the most difficult to process because they can easily become bogged down in scientific disputes and interdepartmental rivalries. In such cases the media will tire of a problem, relegating it to a journalistic limbo where it is considered neither finally retired nor sufficiently topical to be of current public interest.

Mass-mediated environmental discourse

From a topic with no distinct identity of its own, the environment has progressed to the point where it is now an established part of everyday journalism. For example, in 1992, the *Reader's Guide to Periodical Literature* featured a full eight pages of listings pertaining to the environment and the environmental movement. While there has been an upsurge in coverage, there is no single overarching environmental discourse. Instead, the media are the site of multiple outlooks and approaches, some of which are in direct conflict with the others.

At one level, environmental communication is primarily an objectivist scientific discourse. As noted earlier in this chapter, journalists normally view themselves as impartial judges open to conversion only if the scientific proof is seen to be convincing. Scientific claims are reported at face value with relatively little attention to their constructed nature nor to their unknowns and uncertainties[10] (Stocking and Holstein 1993: 202). Journalists have little patience with the thrusts and parries of scientific debate: either a danger exists or it does not.

At the same time, the media routinely lapse into a human interest discourse which 'carries the journalist out of the field of natural science and into the action oriented fields of social movements and politics' (Killingsworth and Palmer 1992: 135). Here, the burden of proof is less exacting. The essence of an environmental problem is more likely to be presented in a single dramatic image: a drum of toxic material, a discarded syringe on the beach, a head of foam on the surface of a trickling stream. Scientific scepticism is replaced by 'common sense'. The emphasis is less on the nature of the conditions which underlie the problem and more on the imputed consequences for people's lives. The narrative is more dramatic, even mythological.

Take for example a recent wire service story (Lawson 1994) on public hearings into a request by a joint Canadian–American venture to convert an unused oil pipeline running through rural Ontario to a natural gas conveyance. Rather than examine the technical, economic and environmental feasibility of the project, the reporter chose to emphasise the participation as an intervenor of Jean Lewington, the widow of an area farmer who had spent thirty years successfully fighting a previous pipeline extension, thereby changing the way utility companies must deal with farmers and their land. This frame was accented by a photograph of Mrs Lewington in front of a barn and a headline which read 'Farm widow refights old pipeline foe'.

Third, the media, especially the business press, have increasingly adopted a discourse which presents the environment as an economic opportunity. The key message here is that environmental adversity can be turned into profit through human ingenuity and industry. Much of this type of coverage is product-oriented, touting a wide variety of 'green' products from the energy-saving house to nuts harvested by indigenous peoples in the rainforests of Brazil. The predominant message is that the entrepreneurial spirit need not be incompatible with ecological values; rather, the two are mutually reinforcing. This optimistic view of the environment has recently been amplified in the rapidly expanding body of stories on the promise and prospects for 'sustainable development'.

Fourth, the media situate the environment as the locus for rancorous conflict. While this environment as conflict package sometimes deals with the wider clash of cosmologies between environmentalists and their opponents, it is more likely to depict these disputes in the same manner as journalists routinely portray industrial relations disputes. That is to say, protesters are implicitly blamed for the disruption of normal commerce, the rationale for their actions is compressed into short sound bites and the background to the conflict is downplayed. The leaders of environmental protest actions are often presented as 'hippies and violent "ecoteurs" armed and ready for monkey wrenching'[11] (Capuzza 1992: 12). An environmental conflict story may shoot to the top of the news agenda if a well-known celebrity arrives on the scene. For example, the protest against the clear-cutting of the old forest on Clayoquot Sound on Vancouver Island was elevated in news value when Robert Kennedy jun. arrived to 'inspect the damage'. Rancorous environmental conflicts are supercharged with symbolic content with both protesters and their opponents likely to use the framing and reasoning devices identified by Gamson and Modigliani (1989).

One consequence is the spill-over of this media discourse into real life ideological battles between environmentalists and their opponents. Thus Dunk (1994) observed that the forest workers in north-western Ontario tend to regard environmentalists as outsiders from 'down south' or from 'big cities', in large part because they uncritically accept the dominant normative structure of the popular media's representation of environmental issues as a confrontation between middle-class, urban-based environmental radicals and local citizens fighting to keep their jobs.

Fifth, the media situate the environment within an apocalyptic narrative. Employing a series of medical metaphors, our planet is depicted as facing a debilitating, perhaps terminal, illness. Overpopulation, loss of biodiversity, rainforest destruction, ozone depletion and global warming are all linked causally to this impending ecological crisis. Despite the caution expressed in scientific media discourse, journalists give considerable news space to the popularised accounts of global threats formulated by Paul Ehrlich (overpopulation, biodiversity loss), Steven Schneider (global warming) and Norman Myers (tropical deforestation) and other prophets of doom. Thus *Time* magazine subtitled its 1989 special issue cover story on the greenhouse effect with the caption, 'Greenhouse gases could create a climatic calamity' (Killingsworth and Palmer 1992: 158).

Finally, the environment is scrutinised through the lens of institutional decision-making. Rather than attributing it a unique status, the environment is treated as another policy area alongside health care, education and social services. The focus here is on regulatory agencies and processes, impending legislation, political personalities (e.g. Al Gore, Maurice Strong) and international fora (United Nations, European Community). Too often this leads to ingrown policy debate between political and scientific elites (Wilkins and Patterson 1990: 21) in which the public is only an incidental bystander.

At any one time, various of these media packages as well as a plethora of individual news frames may compete for dominance. A single environmental event may have multiple shifting frames as it develops. For example, Daley and O'Neill (1991) trace the odyssey of the *Exxon Valdez* oil spill from a disaster narrative (the public as helpless victims, a catastrophe outside human control) through a crime narrative (the captain was culpable) to an environmental narrative (environmentalists contested the statements and practices of industry and government officials). At the same time, attempts to frame a story may fail. In the *Exxon Valdez* case a competing subsistence narrative (the oil spill posed a threat to native Alaskans' way of life) was all but ignored, appearing only in an indigenous publication, the *Tundra Times*. Journalists are thus faced with choosing from an assortment of narratives, languages and viewpoints at the same time as adhering to the formats and structures imposed by traditional journalistic practice.

CONCLUSION

What should be evident from the discussion in this chapter is the considerable extent to which environmental news is socially constructed. In large measure this is a reflection of the rhythms and constraints inherent in the practice of journalism itself. In addition, it reflects the multiple competing claims which newsworkers must routinely sort out in the course of putting together a story. The central difficulty in reporting on the environment has been summed up by Stocking and Leonard in this way:

> The environmental story is one of the most complicated and pressing stories of our time. It involves abstract and probabilistic science, labyrinthine laws, grandstanding politicians, speculative economics and the complex interplay of individuals and societies. Most agree it concerns the very future of life as we know it on the planet. Perhaps

more than most stories it needs careful, longer-than-bite-sized reporting and analysis now.

(1990: 42)

Whether this depth of coverage is realistically possible is an open question which depends on several factors.

First, editors and producers, the newsworkers who effectively set everyday line-ups and assignments, must see environmentalism as more than a transient phenomenon which loses its lustre once it ceases to register strongly in public opinion polls and government agendas. This is less likely to be the case in regions of the country where environmental conflict is endemic because of a natural resource-based economy. Ironically, the one section of the media where environmental coverage has become institutionalised is in the financial pages, where 'green business' is seen as having increasing economic relevance.

Second, environmental issues must be perceived as occupying a distinctive story niche rather than simply overlapping a multitude of existing subject areas – politics, business, agriculture, science and technology, etc. Without a distinctive image, environmental coverage is destined to always remain event-driven and conflict-oriented. At the same time, environmental problems are by their very nature intricately tied in to economic and political structures and policies, making it difficult and sometimes even inadvisable to consider them separately; for example, this is the case with many Third World 'sustainable development' stories. It is thus difficult to balance the need for a distinct environmental specialty beat with the need for a depth of coverage which may reside in other areas of journalistic expertise.

Finally, some way must be found to combine 'muck-raking' or 'exposure journalism' with the longer-term goals of environmental education and policy reform. Investigative reports in the press or on television programmes such as *60 Minutes, Frontline* or *Fifth Estate* may temporarily shock audiences but they do not necessarily result either in a deeper understanding of an issue or in effective regulatory action. Indeed, sometimes there can be a response quite different from that which is desired by activist claims-makers. Fletcher and Stahlbrand (1992:195) cite what occurred in the early part of this century when Upton Sinclair wrote a widely noticed exposure of the exploitation of immigrant workers in the large meat-packing plants of Chicago in his book *The Jungle* (1905):

His dramatic example of a man falling into a machine and being minced with the meat led not to better protection for workers but rather to meat inspection laws, a reform the meat packers wanted to help them compete in European export markets.

In a similar fashion, a recent segment on *60 Minutes* concernin community activist's fight against an incinerator which, she charged, w emitting toxic pollutants evidently resulted in a number of positive business enquiries from other American municipal governments to the waste management company which operated the facility. There must, then, be some blend of story elements which succeeds in raising an alarm in the public arena and then situating this concern within a clearly defined set of goals for environmental reform.

4

SCIENCE AS AN
ENVIRONMENTAL
CLAIMS-MAKING ACTIVITY

It is rare indeed to find an environmental problem which does not have its origins in a body of scientific research. Acid rain, loss of biodiversity, global warming, ozone depletion, desertification and dioxin poisoning are all examples of problems which first began with a set of scientific observations. Ultimately, it is the scientific underpinnings of these environmental problems which lift them above most other social problems that are more dependent on morally based claims (Yearley 1992: 117).

Furthermore, scientific researchers act as 'gatekeepers', screening potential claims for credibility. In 1988, when the British environmental organisation, Ark, mounted a publicity campaign in which they alleged that melting ice-caps due to global warming would raise sea levels five metres in sixty years, thereby covering much of Britain with water, more sober scientific estimates of less than a metre rise quickly discredited the Ark initiative (Pearce 1991: 288–9).

Yet paradoxically, science itself is frequently the target of environmental claims. One notable example of this is the contemporary debate over genetic engineering and its potentially harmful effects in the environment (see Chapter 9). In cases such as this, claims-makers explicitly reject the technical rationality of science in favour of an alternative cultural rationality which appeals to 'folk wisdom, peer groups and traditions' (Krimsky and Plough 1988: 107). Science is pilloried for interfering with the natural order rather than lauded for lending its authority to a claim.

SCIENCE AS A CLAIMS-MAKING ACTIVITY

The profile of science presented so far would seem to suggest that scientific findings reflect the physical reality of the natural world in a relatively straightforward manner. Science would therefore appear to be a search for truth in which the goal is to obtain a clear reflection of nature, as free as possible from any social and subjective influences which might distort the 'facts'.

Yet to the contrary, the assembly of scientific knowledge is highly dependent on a process of claims-making. In this regard, Aronson (1984) has identified two types of knowledge claims made by scientists: cognitive claims and interpretive claims.

Cognitive claims aim to convert experimental observations, hypotheses and theories into publicly accredited factual knowledge. Blakeslee (1994) describes this conversion process as one in which scientists must adeptly stake novel claims while at the same time fitting them into an established research tradition. She gives as an example the process of cognitive claims-making in the the physics journal, *Physical Review Letters*, in which contributors' letters announcing innovations have come to resemble journalistic accounts of scientific findings complete with an arsenal of rhetorical strategies.

Interpretive claims, on the other hand, are designed to establish the broader implications of the research findings for a non-specialist audience. Interpretive claims implicitly ask lay audiences to certify the social utility of the research, and the content of the claim supplies the reason they should do so. For example, in the case of global warming, the cognitive claim is that gases from cars, power plants and factories are creating a greenhouse effect that will boost the temperature significantly over the next seventy-five years or so. The interpretive claim here is that this heating trend is potentially dangerous because, among other things, it will cause havoc with the existing geography of the Earth, flooding some low-lying areas such as the Netherlands and New Orleans and bringing drought to fertile agricultural regions such as the American Midwest.

Not only do scientists make knowledge claims but they also routinely construct 'ignorance claims' (Smithson 1989). This means that researchers highlight 'gaps' in available scientific knowledge in order to make a case for further research funding or, conversely, to retard further policy action on the grounds that not enough hard data exist to justify regulation or legislative activity (Stocking and Holstein 1993). As will be seen in Chapter 7, this was very much the case with industry-sponsored acid rain research in both Britain and the United States, where in-house scientists working for utility companies and those in industry-sponsored institutions emphasised the circumstantial nature of the available data. Aronson outlines three types of interpretive claims which scientists make: technical, cultural and social problem.

Technical interpretive claims-making occurs when researchers act as scientific advisers to industry and government. This often involves the evaluation of risks posed by controversial technologies (nuclear power, genetic engineering), suspected toxic pollutants (dioxin, mercury) and global hazards (ozone depletion, global warming). While in theory scientific advisers are restricted to a narrow technical assessment role, in reality they incorporate their own political agendas and knowledge claims into their interpretations and recommendations.

Salter (1988) uses the term 'mandated science' to refer to the science which is used for the purposes of formulating public policy including studies commissioned by government officials and regulators to aid in their decision-making. Despite an official face of neutrality flowing from scientific expertise, members of expert panels regularly make moral and political claims and choices. These choices are fashioned as much by policy considerations as by scientific norms. For example, a scientific advisory committee dealing with pesticide safety may be equally aware that banning a chemical compound will negatively affect a $500 million industry, while recommending its use could have serious health effects that will only become evident ten years later. This knowledge, Salter observes, affects the committee's recommendations as much as does their technical data, thereby imbuing their activities with a strong interpretive flavour.

Cultural interpretive claims attempt to develop ideological support both for expenditures on scientific research and for the autonomy of science. The media through which the claims are presented are public speeches, articles in popular scientific magazines (*New Scientist, Scientific American*) and on the op-ed pages of influential newspapers (*New York Times, Washington Post, The Times* (London)), testimony before parliamentary enquiries and participation in government–industry committees and panels. In some cases, the receipt of an international scientific prize allows the researcher a unique platform from which to address broader social and political concerns. This is what occurred in Canada when John Polanyi won a Nobel Prize in chemistry and took advantage of the outpouring of public attention to address a raft of issues from government underfunding of universities to nuclear disarmament and peace. In other cases, the threat of a public review of scientific work can mobilise scientists towards making cultural interpretive claims. For example, Krimsky (1979) has demonstrated that the threat of external intervention and control into recombinant DNA molecule research in the 1970s turned American scientists into surprisingly effective lobbyists for scientific autonomy and the freedom of self-regulation.

Social problem interpretive claims assert the existence of a social problem which a particular scientific specialty is uniquely equipped to solve.

Aronson identified three conditions under which scientists are likely to make such claims.

The first is when a new discipline has no foothold in the academic world and therefore must appeal to external constituencies to obtain funding and political support for its work. To a degree, this has been the case for environmental science which has been routinely criticised by many mainstream scientists for doing research that is defensive or of low quality (Rycroft 1991).

The second condition is when enterprising scientists, ever in search of new publicly derived research funds, attempt to show that their existing research work contributes to the solution of a recognised social problem or that it will successfully solve a previously unrecognised problem. This was characteristic of cancer research in the 1970s and Aids research in the 1980s.

Source: *Saturday Evening Post*, 264(5), September/October 1992

A third condition under which social problems claims-making is likely to occur is when scientists are confronted by social movements which seek to restrict their research. In this situation, scientists are compelled to assemble and promote their own set of interpretive claims to either justify why a problem exists and their research should continue or why their research should not be construed as constituting a problem.

Aronson argues that there is a tendency for the first two forms of interpretive claims, technical and cultural, to eventually be transformed or subsumed by the social problem form because what is basically at stake is the social utility of science. That is, researchers recognise that it is better strategy to proactively make a case for the social benefits of their work rather than wait and subsequently have to justify it in an atmosphere of scepticism and budget slashing.

SCIENTIFIC UNCERTAINTY AND THE CONSTRUCTION OF ENVIRONMENTAL PROBLEMS

What particularly opens the door to the creation and contestation of environmental problems is the inability of science to give absolute proof – unequivocal evidence of safety. Instead, scientists are reduced to offering estimates of probability which often vary widely from one to another. This lack of certainty allows claims-makers both within and outside science to assert that the situation is alarming, that the risk is too high and that society should do something about it.

Furthermore, mainstream science and green activists differ fundamentally as to when human intervention is necessary to protect the environment. This difference in perspective is nicely illustrated in a debate which recently took place in the pages of the British science magazine, *New Scientist*.

Brian Wynne, research director of the Centre for Environmental Change at the University of Lancaster, and Sue Mayer, director of science at Greenpeace UK, argue that the decision whether to take official action on environmental risks should be governed by a *precautionary principle*. This states that if there is reason to suspect that a particular substance or practice is endangering the environment then action should be taken immediately even if the evidence is not ironclad. The rationale behind this view is that it will be too late to respond effectively if we wait for a final scientific resolution years down the road. Where the environment is at risk, there is, they argue, 'no clear cut boundary between science and policy' (Wynne and Mayer 1993: 33).

The opposing position is presented by Alex Milne, a consulting chemist who spent thirty-four years working in the paint industry. Milne rejects the precautionary principle, which he labels as one of the central doctrines of 'green science', as entirely the wrong approach. It is worse, he claims, than the legal principle in *Alice in Wonderland*, where the pattern was 'sentence first, verdict afterward'; here it is 'verdict first, trial afterward and no need for evidence' (1993: 37). The precautionary principle, he concludes, has nothing to do with science: it is entirely an administrative and political matter.

A large measure of the disagreement here revolves around how science should be done. In traditional science, a reductionist principle predominates. This means that researchers break down a problem into the smallest number of constituent parts and look at each part separately, controlling as much variation as possible. If you want to look at the effect of a toxic chemical on the breeding pattern of fish, you isolate the fish in an experimental setting, vary the levels of the chemical and record the birth results. By contrast, a cardinal principle of green science is the necessity of looking at the world holistically. Since everything is connected to everything else, it does not make sense to disassemble an ecological web experimentally. For example, immunity is a complex system which is linked to a variety of factors from genetics to environmental pollution to socio-psychological stress. Causation may be indirect or multiple, making it all but invisible to the reductionist perspective of traditional 'good science' (Wynne and Mayer 1993: 34).

In policy terms, good science manifests itself in the form of an *assimilative approach* which purports to define scientifically the capacity of an ecosystem to assimilate pollutants without harm and then licensing industrial discharges within these 'proven' safe limits. What this ignores, environmentalists charge, is the possibility of a chemical interaction among the polluting chemicals which creates a potential for end effects not anticipated by the assimilative model.

As Salter (1988) has observed, quite different sets of criteria are applied depending on the context in which research evidence is evaluated. Conventional science possesses a deeply ingrained capacity to handle ambiguity; indeed, most journal articles routinely end with the caveat 'further research is needed'. By contrast, the burden of proof is stricter when scientists appear before regulatory hearings or in the court-room. Here, legal concepts such as 'reasonable doubt' are prominent – anathema to scientists who are socialised to always couch their conclusions in conditional terms. In this regard, Yearley (1992: 142) points out that scientific expertise depends on elements of judgements and craft skill, informal aspects of science which can be highlighted in a legal or regulatory hearing to make scientific evidence appear like mere opinion. This tendency is even further exaggerated when environmental groups communicate using a moral discourse in a setting where the conventions of a scientific, legal or regulatory discourse predominate. The precautionary principle is a good example of an environmental principle which operates on a different plane of certainty than do societal control institutions.[1]

The crucial dilemma, then, is that social problem interpretive claims which rest on sound scientific evidence are generally more 'robust' than those claims only supported by opinion (Yearley 1992: 76), but there is a fundamental disagreement between environmentalists, scientists, regulators and legalists over what constitutes sound scientific evidence.

Blowers (1993) has observed that scientific evidence is problematic as a basis for environmental policy-making in five ways. First, there is the problem of cause and effect which we have been discussing; this makes it difficult to establish responsibility for the externalities produced by polluting activities. Second, there is the problem of forecasting impacts; for example, the uncertainty about the incidence, distribution, timing and effects of global warming. Third, uncertainty over the consequences of present actions and the risks imposed on future generations may lead to a paralysis of policy or to a tendency to discount the future risks of present action. Sometimes, in fact, another future focused scenario – the crushing burden of a spiralling national debt – may discourage taking bold ameliorative or prophylactic steps in the here and now. Fourth, the frequent absence or sparsity of environmental data not only makes it more difficult to provide sound scientific judgements but it opens the door to manipulation by vested interests who claim that environmentalists have exaggerated the danger. Finally, the often fragile interpretations of environmental science can easily run aground on the shoals of politics where conflicts between interests dominate. This is especially the case where one is dealing with broad speculative ideas such as the Gaia hypothesis rather than narrower, more empirically capturable linkages.

IDENTIFYING ENVIRONMENTAL PROBLEMS AS SCIENTIFIC ISSUES

It is rare indeed to find an environmental problem which pops up overnight with no past legacy of scientific observation and debate. Rather than grow along a linear path, the process by which environmental problems are identified and evolve as scientific issues is characterised by the creation of a pool of knowledge which expands serendipitously in unexpected directions (Kowalok 1993). Individual pieces of data in this pool may be generated through projects which employ the reductionist methods of traditional science, but in the end it is a flash of holistic insight which leads to final understanding.

Despite appearances to the contrary, the basic outline of many environmental problems has been around for a long time. For example, the theory that greenhouse warming is caused by human generated emissions of carbon dioxide has been known for more than a century but the green-house effect was not considered a priority problem until the 1980s (Cline 1992: 13–14). Similarly the term 'acid rain' together with many of its fundamental principles was first introduced by chemist Robert Angus Smith in 1872 but did not emerge as a full-blown scientific problem until the 1970s (see Chapter 7).

What then propels an environmental problem of long standing into a current scientific claim of critical proportions?

First, the real or perceived magnitude of the condition may suddenly rise to 'crisis' proportions. For example, species extinctions have been increasing steadily since 1600 as human settlements have spread across the globe. Recently, however, it has been claimed that we have seriously tipped the balance between the appearance of new species and the extinction of existing ones (Tolba and El-Kholy 1992). At the same time, the loss of old growth forests and plant and animal species captures the attention and concern of conservation biologists and other scientific claims-makers precisely because these natural resources are down to their last twenty, ten or one per cent, making preservation appear more crucial.

Second, new methodologies, research instruments or data banks may allow scientists to come to conclusions which were impossible earlier on. For example, data provided by the European Air Chemistry Network starting in the 1950s allowed Swedish researcher Svante Oden to advance his pioneering theories about acid rain, while James Lovelock's compari-sons of the concentrations of fluorocarbons in the lower atmosphere with annual amounts of industrial production opened the door to chemists Mario Molina and Sherwood Rowland to document the key link between CFC products and ozone destruction (Kowalok 1993).

Third, the holistic character of global ecosystems means that rising scientific and public interest in one environmental problem readily gener-ates interest in another interrelated problem. Thus scientific concern over tropical deforestation has spread well beyond the boundaries of silviculture due in large part to the key role which the loss of tropical forests plays in what are presently the two highest profile global environmental problems: global warming and the loss of biological diversity. Mazur and Lee (1993) illustrate this in schematic fashion, demonstrating how the rise of public concern over the problem of the global environment is actually a weaving together of several strands of concern over specific problems, each of which has arisen at a different point in time. This synergy is not, of course, always readily apparent and scientific entrepreneurs may need to explicitly estab-lish the relevance of one issue for another.

Fourth, the establishment of official research programmes, centres and networks may create a hothouse in which research into an environmental problem may be successfully nurtured, even if this is not the original intention. For example, the decision in December 1979 by the Council of the European Community to establish a multiannual research programme in the field of climatology was taken in part because of concern about what was essentially a regional problem – the 1976 drought which affected some African and European areas. Once in place, this programme became both the focus of foundation-building research on the physico-chemical processes related to the increasing concentrations of greenhouse gases in the atmosphere and a source from which scientific findings and terms such as 'greenhouse effect' and 'climate change' circulated outwards into EC policy-making circles (Liberatore 1992).

In all of this the identification and characterisation of environmental threats is highly dependent upon a network of international scientific conferences and collaboration (Kowalok 1993: 36–7). Not only does this permit researchers to learn new methodological techniques or to find the missing pieces in their own puzzles but it helps build their confidence that they are not alone, an especially important shot of morale boosting when a theory seems radically new and controversial. This was very much the case with the groundbreaking research on the acid rain problem where Canadian and American researchers did not fully appreciate the global relevance of their own findings until they came face to face with similar findings from Scandinavia as presented by Oden on his 1971 lecture tour of North America (Cowling 1982).

COMING OUT: COMMUNICATING NEW ENVIRONMENTAL PROBLEMS TO THE NON-SCIENTIFIC WORLD

The transition from cognitive to interpretive scientific environmental claim is comparable to a 'coming out' ceremony in which the ingenue makes a public representation of identity. At some point, the circulation of information around an essentially closed scientific loop is interrupted and the urgency and salience of a problem is shared with the outside world.

One common way of doing this is to convene a public forum at which a mixture of scientists, environmentalists and administrators jointly address the various dimensions of the problem in the full glare of the media spotlight. Alternatively, a claim may be articulated at a congressional or parliamentary hearing where media coverage is usually assured. For example, the 1981 US Congressional testimony by Peter Raven and Edward Wilson was important in establishing the economic utility of preserving endangered species of insects such as the butterfly or the honey-bee, particularly for the development of new crops, drugs and renewable energy sources (Kellert 1986). Similarly, the ozone depletion issue in Britain was not effectively launched until parliamentary hearings were held in early summer 1988; strong representations were made in both houses of Parliament to the effect that the United Kingdom must become a world leader in the drive to protect the ozone layer (Benedick 1991). A third channel for the public dissemination of newly constructed scientific environmental problems is a scholarly conference at which reporters from major newspapers are present in their search for 'blockbuster' theories. This is what occurred in September 1974, when the *New York Times* picked up on a delivered paper dealing with the threat of CFCs to the ozone layer; the *Times* article 'signaled the beginning of public concern over CFCs and their use in aerosol cans and refrigerators' (Kowalok 1993: 19).

In other cases, however, this process is short-circuited when scientific entrepreneurs go directly to the media. Svante Oden, the Swedish soil scientist who first proclaimed the theory of acid rain, published an account in the Stockholm newspaper *Dagens Nyheter* a year before he published in a scientific journal and five years before the issue arose at the 1972 UN Conference on the Human Environment. Similarly, in Germany, biochemist Bernhard Ulrich's hypothesis that huge tracts of German forests would be dead within five years due to damage from acid rain was presented as established fact in an article in *Der Spiegel*, a mass circulation periodical, provoking widespread national alarm (see Chapter 7).

How effective one channel is compared to another depends on a number of factors. If there is no consensus among scientists themselves and strident opposition from industry, a more individual approach may work best. Despite periodic attempts to raise the issue, the problem of pesticide poisoning in the United States was being effectively suppressed[2] until Rachel Carson published her indictment in *Silent Spring*. Subsequently, a number of scientists came forward in her defence and the problem was legitimated when, in May 1963, a special panel of the President's Scientific Advisory Committee released a report which was critical of the pesticide industry. On the other hand, jumping the gun before scientific consensus has been established may succeed in capturing media and public attention but at the risk of bringing peer censure by fellow scientists. This is what occurred in 1988 when James Hansen, director of the NASA Institute for Space Studies, testified before a US Senate committee that summer heatwaves such as that which was being experienced at the time were directly attributable to the greenhouse effect. This norm within science against premature revelation has no doubt been strengthened as a result of the 'hoax' over cold fusion in which the researchers announced their findings at a press conference in Utah prior to subjecting them to peer review.

SCIENCE AND ENVIRONMENTAL POLICY-MAKING

In order for a scientific issue to become policy it must be translated into something which is 'treatable'. As a result, at the policy formulation stage the contribution of natural scientists usually diminishes while the role of socioeconomic and technical experts grows. For example, Liberatore (1992) found that while natural science findings still played an important role in the international debate on global warming, it was the input of economists, policy analysts and energy technology experts which was crucial in shaping the nature of the European Community response.

The relationship between science and policy-making has been captured most adequately by political scientists using two concepts: epistemic communities and policy windows.

Epistemic communities

Haas has described the contribution of 'epistemic communities' as critical in achieving international cooperative agreements on environmental issues. Epistemic communities are 'transnationally organized networks of knowledge based communities'; that is, internationally linked groups of specialists who offer technical advice to political decision-makers.

What gives them a key role in a process usually closed to non-politicians is the uncertain nature of environmental problems. Political leaders may be highly skilled in negotiating trade pacts or military treaties but they feel at a distinct disadvantage in dealing with planet-threatening conditions relating to atmospheric shifts or chemical overloads. Under such circumstances information is at a premium as a strategic resource, and politicians, in order to reduce such uncertainty, 'may be expected to look for individuals who are able to provide authoritative advice on whom to pin the blame for a policy failure or simply as a stop-gap measure to appease public clamour for action' (Haas 1992: 42).

Epistemic communities, Haas contends, are not only bound together by a common technical expertise but they also share a number of causal and principled beliefs. In the case of environmental issues, these communities of knowledge have been largely comprised of ecologists who share a common belief in the need for a holistic analysis – a view which carries over to the policy advice that they give. This was characteristic of an epistemic community of ecologists and marine scientists who spear-headed intergovernmental efforts in the 1980s to control pollution in the Mediterranean Sea (Haas 1990).

An epistemic community has the capacity to be influential both in defining the dimensions of a problem and in identifying likely solutions. For example, Haas demonstrates how a transnational epistemic community of atmospheric scientists was successful in influencing the negotiations which led to the signing of the Montreal Protocol on the protection of the ozone layer in September 1987 by 'bounding discussions on the broad array of substances to be covered and the rapidity of regulations' (Haas 1992: 49). Once the epistemic community has laid out the basic parameters of the settlement, it is then up to the political leaders to decide what compromises have to be made in order to obtain agreement.

It should be noted, however, that not all political analysts agree with Haas's elevation of scientific coalitions to a central place in the environmental decision-making process. Haas's model is said to break down in the degree of autonomous power accorded to the epistemic community. That is, scientific coalitions can use their resources to highlight a problem but they must enlist political leaders from their individual nations to have a real impact on environmental treaty negotiations. These leaders may find it advantageous to engage in international problem-solving but ultimately they are guided by domestic political considerations (Susskind 1994: 74–5).

Individual governments depend on the technical expertise built up by environmental movement organisations such as Friends of the Earth, Greenpeace and Pollution Probe. In recent years, these groups have devoted considerable resources to building up their own in-house research capabilities, hiring scores of bright, young, idealistic Ph.D.s fresh out of graduate school. In addition, conservation and environmental organisations typically have scientific advisory committees and call upon the voluntary support of university scientists and civil servants who are scientists (Yearley 1992: 126). As a result, there is a synergy between organisations and official policy-makers who find the knowledge and information produced by Greenpeace and others to be of considerable strategic value in staking out their position in public arena debates over environmental issues (Eyerman and Jamison 1989; Lowe and Goyder 1983).

While epistemic communities may be international in scope, the centre of gravity for scientific claims-making on specific issues tends to reside in a specific nation. For example, it was US scientific leadership which propelled the ozone depletion problem into global prominence while Swedish (and Norwegian) research on acid rain was vital in elevating that issue to problem status. In the former case, a critical infrastructure clearly

existed as the result of the space programme and the pre-eminence it gave
to the United States in researching the stratospheric sciences. This was
particularly located in two government agencies – NASA (National
Aeronautics and Space Adminstration) and NOAA (National Oceanic and
Atmospheric Administration) – as well as in the graduate faculties of major
American universities (California, Harvard, Michigan). When researchers
at these institutions voiced concern about events in the stratosphere, the
site of the ozone problem, the media and the general public as well as
political leaders tended to pay attention (Benedick 1991). In the case of
acid rain, the forests and lakes were seen as a vital component of the
Swedish economy and recreational life and therefore were accorded high
research priority. When the transnational origins of acid precipitation
became obvious in the research data reported by Oden and others, the
Swedish government did not hesitate to aggressively present these findings
at the 1972 Stockholm Conference.

Policy windows

Another political science model which can be used to link science and
domestic environmental policy-making is Kingdon's 'garbage can' model.
Adapted from a model of organisational choice developed by James March
and his colleagues, this proposes the operation of three major process
streams in government agenda setting: (1) problem recognition; (2) the
formation and refining of policy proposals; and (3) politics. These three
streams usually develop and operate largely independently of one another.
However, at critical times the three streams may come together or 'couple'.
Kingdon describes this as the opening of a 'policy window' and attributes
the main responsibility for this action to 'policy entrepreneurs' within the
political system. Individual entrepreneurs do not open the window but
they take advantage of the opportunity once it has occurred. At key
junctures, then, solutions become joined to problems and both are joined
to favourable political forces.

Hart and Victor (1993) have recently employed Kingdon's model to
explore the role of scientific elites in influencing American policy on
climate change for the years 1957–74. In their interpretation, science,
policy and politics evolve in separate unconnected streams creating both
solutions in search of problems and political problems in search of solutions.
Scientific elites, assuming the entrepreneurial role, play a central role in
identifying policy windows and seizing advantage of them.

This is what occurred in the United States in the 1970s. For the better
part of twenty years, two interesting scientific discourses relating to the

climate had been meandering along, attracting some support but unable to really get moving in terms of either funding or public recognition. These were the 'carbon cycle discourse' which addressed the question of whether and why atmospheric concentrations of carbon dioxide (CO_2) were increasing and the 'atmospheric modelling' discourse which asked what would happen to the climate if higher concentrations of CO_2 were reached. The former discourse was coordinated by an oceanographer, Roger Revelle, while the latter was promoted by John von Neumann, the father of scientific computing.

In the early 1970s, the rise of the American environmental movement created a policy window which these elite scientists successfully exploited in order to mobilise financial and political support and raise public awareness. Hart and Victor (1993: 661) describe this as a synergistic relationship in which scientific findings such as those relating to the greenhouse effect 'catalysed the rebirth of environmentalism' while at the same time environmentalism 'acted as a midwife for new scientific agendas – legitimating them and providing constituencies for their results'. Especially influential in linking the two research streams was Carroll Wilson, an MIT management professor, who was the guiding spirit behind the publication in 1970 of a report, entitled *Study of Critical Environmental Problems*, which was explicitly interdisciplinary and environmentalist in tone.

Hart and Victor (1993: 668) emphasise that very little new scientific information about the prospects of global warming was produced between the late 1960s and the early 1970s. Rather, what was different was that the two lines of research were brought together in a new, redefined, scientific agenda which was then successfully sold to political decision-makers and to the news media as a global environmental 'pollution' problem.

SCIENTIFIC ROLES IN ENVIRONMENTAL PROBLEM-SOLVING

Susskind (1994) has proposed five primary 'roles' which are played by scientific advisers in the environmental policy-making process: trend spotters, theory builders, theory testers, science communicators and applied policy analysts. These roles frequently overlap but each has its own tasks and agendas.

Trend spotters are scientists who are the first to detect changes in ecological patterns and to correctly understand their significance. Occasionally, the trend spotter may be a lone scientist who observes some important pattern in the microecology of the pond or marsh and is able to extrapolate this on to the larger environmental canvas. More common, however, are trend spotters who are part of a scientific team which is engaged in gathering and analysing longitudinal data such as that assembled from the LANDSAT satellite or from the European Air Chemistry Network.

Theory builders try to explain the causes for the changes that the trend spotters identify. They are inclined to engage in model building, both to fit explanations to past circumstances and to predict future effects.

Theory testers critically scrutinise the models suggested by theory builders. Using pilot tests or controlled experiments, they attempt to ascertain whether the hypotheses and propositions generated by the model can be empirically proven.

Science communicators attempt to translate difficult-to-decipher data into terms that the public at large can understand. They are key players in the 'coming out' process which was discussed in an earlier section of this chapter. Some communicators such as Edward Wilson are eminent scientists who feel a strong moral responsibility to bring the fruits of their research to the public. Others, for example, the Canadian geneticist and broadcaster David Suzuki, are researchers who have made a conscious decision to spend their life popularising science and carrying the ecological message to a wider audience.

Applied policy analysts act as consultants to political decision-makers, converting scientific findings into policy recommendations. They play a prominent role in the formulation of environmental treaties because they take what is often abstract scientific information and recast it in terms which are amenable to legislation or to international agreements.

Each of the five types of scientists may contribute throughout the environmental problem-solving process but there is a considerable degree of specialisation; that is, trend spotters and theory testers are usually more prominent during the fact-finding stages while science communicators and policy analysts play key roles during the negotiation/bargaining period (Susskind 1994: 77). In terms of the three key tasks in constructing environmental problems discussed in Chapter 2, trend spotters and theory testers can be said to characterise the 'assembling' process, communicators in 'presenting' an issue and applied policy analysts in 'contesting' an environmental claim (see Table 1, page 42).

REGULATORY SCIENCE AND THE ENVIRONMENT

One important arena in which environmental science interacts with politics is in the regulatory process. The 'regulatory science' which is found here differs from conventional research science in a number of ways (Jasanoff 1990). First, it is more often done at the margins of existing knowledge where fixed guidelines for evaluation may often be unavailable. Second, it usually involves a greater degree of 'knowledge synthesis' than does research science which puts a greater emphasis on the originality of findings. Third, science-based regulation requires a hefty element of 'prediction' especially with regards to risk creation.

Jasanoff (1990:230) argues that a negotiated and constructed model of scientific knowledge 'closely captures the realities of regulatory science'. Rather than encouraging an adversarial process, regulatory agencies seek scientific input into their decisions as a means of legitimation. This often takes the form of an ongoing scientific advisory committee. Jasanoff reviews a number of cases in which such advisory boards played a key role in decision-making at the Environmental Protection Agency (EPA) in the United States. In the case of air pollution, the relationship between the EPA and the Clean Air Scientific Advisory Committee (CASAC) was initially rocky but after extensive negotiation was transformed into a fundamentally sympathetic orientation. Similarly, despite problems during the Reagan era, the EPA's agency-wide Scientific Advisory Board (SAB) was able to maintain a respected and autonomous position, in large part because it focused on issues pertaining to scientific assessment while leaving rule-making activities to the agency proper.

In this negotiated model of regulatory science, Jasanoff contends, there can be no 'perfect, objectively verifiable truth', only a 'serviceable truth' which balances scientific acceptability with the public interest. In this context, scientific reality is clearly socially constructed so as to conform to a societal mean. However, in circumstances where sharply conflicting constructions of science land at the feet of a scientific advisory committee, reconciliation can often be most difficult. This is what has occurred in various regulatory controversies involving agricultural pesticides where scientific evidence has been especially difficult to establish while public concern has been high. In these situations, the debate over the 'precautionary principle' which we surveyed earlier in this chapter rears its head, with scientific advisers opting for the traditional reductionist position while agency staff are more sensitive to the public pressure to act sooner rather than later. Where this occurs, the risk debate can easily shift to the arenas of the media and politics where it will continue under a different set of ground rules from those encountered in the regulatory setting (Jasanoff 1990: 151).

5

CONSTRUCTING ENVIRONMENTAL RISKS

Hot dogs have long been a symbol of American culture, a mainstay of countless social and community events from baseball games to backyard barbeques and school picnics. Despite questionable nutritional value, they are especially popular with families because they are inexpensive, easy to prepare and children ask for them.

Now hot dogs have joined the growing list of food products which have been related to an elevated risk of cancer. A recent American research study, widely reported in the media, linked moderately high hot dog consumption with an increased risk of childhood leukaemia. Time will tell whether these findings have any long-term impact or whether they will be quickly forgotten. Nevertheless, for a short while, at least, hot dog sales are likely to dip in some places as parents exercise caution in their diet choices.

To a large extent, this episode is characteristic of how individuals in contemporary society engage in the processes of risk perception and assessment. Typically, we hear a brief item on the radio or see it in a newspaper or magazine, it comes from a seemingly reputable scientific source and it taps into an existing well of concern about our health or the safety of our family. This is true not only for food and lifestyle choices but also for risks related to technology and the natural environment.

Until recently, the published literature on risk almost uniformly reflected the belief that risks could be 'objectively' determined, that this determination was exclusively the province of engineers, scientists and other experts and that any failure on the part of ordinary citizens to fully accept this was considered irrational. Risk assessment was thus conceived of as a technical activity where results were to be formulated in terms of 'probabilities'. There was even an emerging category of specialists – what Dietz and Rycroft (1987) have termed the 'risk professionals' – who make it their business to work out new methods of risk analysis.

RISK AND CULTURE

The first notable challenge to this position came from a British social anthropologist, Mary Douglas, and an American political scientist, Aaron Wildavsky, who published a provocative book in 1982 entitled *Risk and Culture: An Essay on the Selection of Technological and Environmental Dangers*.

Risk and Culture asks two simple but fundamental questions. Why do people emphasise certain risks while ignoring others? And, more specifically, why have so many people in our society singled out pollution as a source of concern? The answers, Douglas and Wildavsky insist, are embedded in culture.

In their view, social relations are organised into three major patterns: the individualist, the hierarchical and the egalitarian. Individualist arrangements are based on the laws of the market-place while hierarchical relations are epitomised by government bureaucracies. Egalitarian groups are aligned in a 'border zone' on the margins of power at the political economic centre of society where the other two modes of social organisation are normally located.

Egalitarian groups have a cosmology or world-view which is more or less equivalent to the 'New Ecological Paradigm' discussed by Catton and Dunlap, Cotgrove and others. Unbridled economic growth is frowned upon, the authority of science is questioned and our boundless faith in technology is declared unwise.

Douglas and Wildavsky's central thesis is that the perception of risk varies considerably across these three forms of social organisation. Market individualists are primarily concerned with the upswing/downturn of the stock market, hierarchists with threats to domestic law and order or the international balance of power and egalitarians with the state of the environment. This leads them to conclude that the selection of risks for public attention is based less on the depth of scientific evidence or on the likelihood of danger but rather according to whose voice predominates in the evaluation and processing of information about hazardous issues.

In this view, the public perception of risk and its acceptable levels are 'collective constructs' (Douglas and Wildavsky 1982: 186). No one definition of risk is inherently correct; all are biased since competing claims, each arising from different cultures, 'confer different meanings on situations, events, objects, and especially relationships' (Dake 1992: 27).

Unfortunately, at this point, Douglas and Wildavsky's cultural theory of risk slips off the rails on to spongier ground. Environmental egalitarians, they suggest, are the secular equivalents of religious sects such as the Anabaptists, the Hutterites and the Amish. Obsessed with doctrinal purity and the need for unquestioned internal loyalty, sectarians are seen as having to create an image of threatening evil on a cosmic scale. It is therefore necessary and 'functional' for environmental sectarians such as those found in Friends of the Earth to constantly identify new risks from nuclear winter to global warming. Each new crisis is chosen, they maintain, 'out of the necessity of maintaining cohesion by validating both the sect's distrust of the center and its apocalyptic expectations' (Rubin 1994: 236). This explains why they turn their back on local causes in favour of global issues so vast in scale as to warrant invoking a sense of general doom. Pollution and other environmental risks are therefore used by these sectarian challengers as a means of holding their membership together and for attacking the establishment groups of the centre which they oppose (Covello and Johnson 1987: x).

Risk and Culture has provoked much interest and a torrent of criticism. Much of the latter focused on the claim that environmentalists mobilise primarily for solidary rather than purposive reasons. That is, rather than view environmentalists as part of a moral response to a very real societal crisis, they have chosen to treat environmental risks as merely bogymen which serve the same purpose as certain food prohibitions among tribal peoples. Environmentalists, therefore, are not regarded as rational actors but rather as 'true believers' open to manipulation by ecological prophets such as Barry Commoner, David Brower and Edward Abbey.

Karl Dake, a member of the Douglas–Wildavsky research circle,[1] has insisted that this criticism is overstated and that the cultural school of risk never meant to imply that perceived dangers are simply manufactured:

> People do die; plant and animal species are lost forever. Rather, the point is that world views provide powerful cultural lenses, magnifying one danger, obscuring another threat, selecting others for minimal attention or even disregard.
>
> (Dake 1992:33)

Douglas and Wildavsky are less accommodating, however, insisting that knowledge about risk and the environment is 'not so much like a building eventually to be finished but more like an airport always under construction' (1982: 192). It is fruitless, they claim, for the social analyst to try to assess whether the risk under discussion is real or not; what matters is that the debate keeps going with 'new definitions and solutions'. Rubin (1994: 238–9) totally rejects this relativism, arguing that public policy considerations require that we know *definitively* whether risks such as those arising from global warming or ozone depletion are merely foils for the apocalyptic needs of sectarian organisations or genuine threats which must be dealt with. While Rubin's point is well taken, the ambiguity of many contemporary risks makes it difficult to achieve the certainty which he would like to see. Even if we reject Douglas and Wildavsky's absolute relativism, nevertheless the by now widely accepted argument which they make about the subjective and imprecise nature of scientific findings militates against the infallibility of expert opinion. As a society, we still have to make social judgements about the magnitude of risk, although scientific evidence can be one helpful source of information in making these decisions.

SOCIOLOGICAL PERSPECTIVES ON RISK

Sociologists of risk generally adopt a more moderate position than that of Douglas and Wildavsky, insisting that while risk is certainly a sociocultural construct, it cannot be confined to perceptions and social constructions alone. Rather, technical risk analyses are an integral part of the social processing of risk (Renn 1992).

Dietz *et al.* (forthcoming) have observed that the main currents in the sociology of risk have followed three separate but complementary directions which are bound together by an underlying emphasis on the social context in which individual and institutional decisions about risks are made.

First, sociologists have been concerned with the question of how perceptions of risk differ across populations facing different life chances and whether the framing of choices stems primarily from power differences among social actors. Thus Heimer (1988) points out that the residents of Love Canal saw the risks from chemical dumps differently from executives of the Hooker Chemical Company and from bureaucrats in the state government and various state agencies which deal with public health and the environment. Similarly, workers and bosses see environmental health risks in the workplace in a different light. To a certain extent, this issue overlaps the social distribution of risk, although the emphasis here is on how social location affects the perception of risk rather than on how it alters the likelihood of being exposed to hazardous conditions.

Second, sociologists of risk have proposed a model that reconceptualises the problem of risk perception by taking into account the social context in which human perceptions are formed. That is, individual perception is powerfully affected by a panoply of primary (friends, family, co-workers) and secondary (public figures, mass media) influences which function as filters in the diffusion of information in the community. This is captured in the concept of 'personal influence' which was central in the mass communication research of the 1950s and 1960s (see Katz and Lazarsfeld 1955).

Third, risks, especially those of technological origin, have been conceptualised as components of complex organisational systems. This is exemplified by Perrow's (1984) well-known analysis of 'normal accidents' in which an estimated probability of failure is built right into the design of technologies with high catastrophic potential. Once implemented, however, such systems severely limit any further human ability to manipulate risks since the source of the risk is now located in the organisation itself (Clarke and Short 1993).

Renn (1992) has further classified the sociological approaches along two dimensions: (1) individualistic versus structural, and (2) objective versus constructionist. The first dimension asks whether the approach in question maintains that the risk can be explained by individual intentions or by organisational arrangements. Objectivist concepts imply that risks and their manifestations are real, observable events while constructionist concepts claim that they are *social artifacts* fabricated by social groups or institutions. According to this taxonomy, the first two currents of risk research identified by Dietz and his colleagues tend to be individualistic/constructionist while the third is structural/objective. Notable by its absence is a 'social constructionist' perspective which Renn describes as an approach which 'treats risk as social constructs that are determined by structural forces on society' (1992: 71).

SOCIAL DEFINITION OF RISK

Hilgartner (1992) has argued that the constructionist perspective must begin by examining the conceptual structure of social definitions of risk. Such definitions, he maintains, include three major conceptual elements: an *object* deemed to pose the risk; a putative *harm*; and a *linkage* alleging some causal relationship between the object and the harm.

To assume that objects are simply waiting in the world to be perceived or defined as risky is 'fundamentally unsociological' (Hilgartner 1992: 41). Rather, an initial phase of risk construction consists of isolating and targeting the object(s) which constitute(s) the primary source of a risk.

In the late 1980s, the lakeside urban neighbourhood in which my family and I currently reside was designated by the municipal public works department to receive a pair of 'sewage detention tanks', one to be installed in Kew Gardens, a multi-use community park, the other on the beach adjacent to the boardwalk. The problem, we were told, was effluent from the City's storm sewer system which flowed into Lake Ontario and made it too polluted with faecal coliform bacteria to allow swimming. According to studies conducted by an engineering firm engaged by the City, there were two primary sources from which the faecal coliform pollution originated: human faeces contained in combined sewer overflow[2] and animal excrement which had been swept along with rainwater into the storm sewers.

Our residents' association which first learned of the project when one member came across the publication of a statutory notice buried in the pages of a local daily newspaper, at first expressed concern on the grounds of the disruption which construction would bring to the park and the beach, both of which are heavily used. However, in the course of researching the proposal and meeting with other residents, we began to realise that, in fact, the source of the risk probably did not reside primarily in the storm-water but in effluent which was being dumped into the lake from the main sewage treatment plant located just to the west of our neighbourhood. We learned that, due to insufficient capacity, operators at this plant routinely opened the sea-wall gates just before it began to rain and released untreated or partially treated sewage into the lake at levels 10,000 times that at which the beaches were declared unsafe for swimming and closed. On one day out of three the lake currents reversed direction, sending this effluent towards our beaches. Immediately after a public meeting one night, a retired operator at the drinking-water filtration plant located at the eastern fringe of the neighbourhood told me that he used to routinely receive a telephone call from his equivalent at the sewage treatment plant advising that in advance of rain they were opening the gates and that he should raise the chlorine levels – a tip-off that the coliform pollution was migrating along the near shore area in a kind of bathtub ring pattern. We did not know it at the time but a somewhat similar situation occurs regularly in Sydney, Australia, where the ageing sewage system which pumps sewage out to sea is designed to overflow into storm sewers during periods of heavy rainfall so as not to clog up already overloaded treatment tanks (Perry 1994: WS-4).

What happened here was that residents opposed to the sewage detention tanks developed an alternative definition of the 'risk object'. At public meetings, at City Hall and at a special hearing before an Environmental Assessment Advisory Committee appointed by the provincial Minister of the Environment to consider whether to grant our request for a 'bump up' (i.e. from a routine class environmental assessment to a more formal and rigorous individual environmental assessment), we actively contested the official designation of the object deemed to be risky and presented our claim (unsuccessfully) that the main sewage treatment plant was the villain instead.

The second element in the social definition of risk involves the process of defining harm. Once again, this is not as obvious as it may seem. For example, forest fires are commonly thought to wreak a path of destruction but ecologists contend that in nature they serve a useful function in woodland renewal. Offshore oil-drilling platforms are assumed to pollute the waters surrounding them but marine biologists have found that they also spawn a whole new microecology at their base. Some environmentalists in the United States have recently campaigned to reduce the allowable levels of the trace mineral selenium which can be added to animal rations on the basis that it leaves toxic residues, but representatives of the feed industry maintain that selenium additives are a boon to the environment because they reduce the amount of feed consumed thus saving on energy.

In each of these cases, the very definition of what harm ensues from a particular object or action is contested, sparking a variety of claims and counter-claims, despite the fact that there is mutual agreement as to the risk object (forest fires, offshore oil drilling, selenium as a feed additive). Risk claims may frequently conflict on ideational grounds. Thus, a river diversion project which provides irrigation water for local farmers (a human benefit) may result in the destruction of a fragile ecosystem of fish, birds, insects, etc. (a biological harm). Similarly, the road salt which is deemed so vital to coping with the harsh winter in parts of Canada and the Northern United States has been declared by scientists as harmful to the lakes, rivers and streams where it is eventually deposited. Conversely, environmental initiatives which are declared to be of ecological benefit may result in problems for human constituencies. For example, the protection of wolves is advocated by some wildlife preservationists but it is keenly opposed by ranchers who fear the loss of livestock crucial to their economic survival. With consensus impossible, the central basis of contestation becomes the presence or absence of harm which is generated by a risk object.

A third component of the social construction of risk consists of the linkages alleging some form of causation between the risk object and the potential harm. Hilgartner (1992: 42) observes that constructing these linkages is always problematic because a risk can be attributed to multiple objects. Indeed, the 'laws' of ecology encourage this, since all things are regarded as being interdependent. This is further complicated by the fact that the full extent of the risk may not be known until many years later. For example, a recent report by a Minnesota radio station suggests that a 1953 US Army test, in which clouds of zinc cadmium sulphide, a suspected carcinogen, were sprayed aerially over Minneapolis dozens of times may have caused an unusual number of stillbirths and miscarriages; these problems have shown up particularly often in former students of a public elementary school which was one of the spray sites forty years ago (*New York Times* 1994). The effects may sometimes be more immediate but it takes years for claims-makers to assemble them into a publicly acknowledged form. This was the case with a raft of health-related ailments among military veterans of the Gulf War. Even though symptoms began soon after their return, it is only now that public reports of a 'Gulf War syndrome' are beginning to permeate the mainstream media and to be framed in terms of deleterious environmental agents in the war zone.

Much of the discourse over the social construction of risk takes place on this terrain. The situation is further complicated by the existence of multiple conflicting layers of proof: legal scientific, moral.

The burden of legal proof is the most onerous, since it cannot leave any room for 'reasonable doubt'. The caveats which are standard in scientific studies (e.g. 'the data are suggestive but require further research') do not stand up in court. Nor usually do anecdotal or clinical evidence.[3] As environmentalists have discovered, judges are often loath to break any new ground by acting to prevent a problem before it happens.

Scientific proof is easier to come by, but nevertheless is a slave to statistical significance levels. It is also notoriously fickle, its authority intact only until the next disconfirming study appears. The harmful effects of acid rain on lakes and forests, for example, is constantly being discovered, debunked and rediscovered (see Chapter 7). The scientific layer of proof can be subdivided into two standards: a standard drawn from pure science in which action is not recommended until correlations weigh in at the 95 per cent confidence level, and a standard utilised by the medical disciplines in which action may be taken before significance is reached if the evidence points towards a serious health problem.

Collingridge and Reeve (1986) have demonstrated the clash between these two versions of scientific evidence in the debate over the health effects on children of lead from vehicle exhausts. In the US it haunted the conflict between the EPA, which supported the removal of lead from gasoline (petrol) on the basis of broad differences in blood lead levels among urban and suburban populations, and the Ethyl Corporation, a major manufacturer of lead additives, which argued that the link between blood and air levels remained statistically unproven. In the UK, the difficulties arose in the early 1980s between the government-sponsored 'Lawther Report' which rejected all laboratory animal and biochemical studies as irrelevant to understanding the medical effects of lead on humans and the report entitled *Lead or Health* by the environmental group, the Conservation Society, which argued the contrary: 'Moral proofs are most easily manufactured but are heavily dependent upon the mobilisation of public opinion in order to make an impact.'

The use of moral proofs allows the formation of attitudes or opinions about a risk issue even if the scientific or legal layers of proof indicate a degree of uncertainty or ambiguity. For example, animal rightists have never been able to conclusively prove scientifically that animals 'suffer', so they have adopted the alternative strategy of trying to demonstrate ethically that this is the case, drawing in particular on the work of the philosopher Peter Singer. Similarly, the case against the biological engineering of plants and animals is not empirically strong (no genetically altered fruits have thus far performed like the protagonist in the Roald Dahl story, *James and the Giant Peach*) but the moral case against interfering with nature is more impressive. Such moralisation, however, tends to polarise positions on risk policies, making compromises more difficult (Renn 1992: 192).

Unlike the legal and the scientific, the most effective moral proofs are often those which follow a simple line of reasoning. Consider, for example, the nature of the argument presented by 'Kapox' – labelled by the South American press as the 'Tarzan of the Amazon'. Kapox, who engages in long-distance swims through the Amazon region to publicise the state of pollution of the river and the destruction of the surrounding rainforest, does not base his appeal on a sophisticated reasoning about the need to protect biodiversity. Rather, he preaches a simple, obvious, moral message: as the largest river in the world concentrating a fifth of the planet's fresh water, the Amazon deserves respect (Suzuki 1994a).

ARENAS OF RISK CONSTRUCTION

As powerful as Kapox's appeal may be, it is unlikely to directly influence collective risk decisions or policies. Instead, social definitions of environmental risk must be followed up by political actions designed to mitigate or control the risk which has been identified. Building on the work of Hilgartner and Bosk (1988), Renn (1992) argues that political debates about risk issues are invariably conducted within the framework of 'social arenas'.

The term *social arena* is a metaphor to describe the political setting in which actors direct their claims to decision-makers in hopes of influencing the policy process. Renn conceives of several different (theatre) 'stages' sharing this arena: legislative, administrative, judicial, scientific and mass media. While both traditional and unorthodox action strategies are permitted, these arenas are nevertheless regulated by an established repertoire of norms. For example, illegal direct action such as that advocated by Earth

First, the American renegade environmental group, violates this protocol. The code is, in fact, a combination of formal and informal rules usually monitored and coordinated by some type of enforcement or regulatory agency such as the EPA in the United States and the Department of the Environment (D.o.E) in Britain.

The concept of the social arena combines elements from the organisation–environment perspective in the field of complex organisations, Goffman's dramaturgical model of social relations and the symbolic models of politics as developed by Murray Edelman (1964, 1977) cemented together by a social constructionist compound. As formulated by Renn, it also stresses the mobilisation of social resources as discussed by the McCarthy–Zald school within the resource mobilisation perspective on social movements. Renn seems unaware of the parallels but the social arena concepts which he uses also echo some recent research on international environmental diplomacy, notably Haas's (1990, 1992) conceptualisation of 'epistemic communities' (see Chapter 4).

While some elements of risk construction may occur in the public domain beyond their parameters, the most important action takes place in arenas that are populated by communities of specialised professionals: scientists, engineers, lawyers, medical doctors, government officials, corporate managers, political operatives, etc. (Hilgartner 1992: 52). Such technical experts are the chief constructors of risk, setting an agenda which often includes direct public input only during the latter stages of consideration. Hilgartner and Bosk (1988) note that these 'communities of operatives' often function in a symbiotic fashion, the operatives in each arena feeding the activities of operatives in the others. Environmental operatives (environmental groups, industry lobbyists and public relations personnel, political champions, environmental lawyers, journalists and bureaucrats) are notable examples of this; by virtue of their activities they both generate work for one another and raise the prominence of the environment as a source of social problems.

Within the social arena of risk, the process of defining what is acceptable and what is not is often rooted in negotiations among several or multiple organisations seeking to structure relations among themselves. Clarke (1988) illustrates this in his analysis of an office building fire in Binghamton, New York, which left a legacy of toxic chemical contamination. In this case, three governmental agencies – the state health department, the county health department and the state maintenance organisation – collectively vied for suzerainty in determining how risky the situation was thought to be. In such cases, Clarke argues, the institutional assessment of risk is a claims-making activity in which corporations and government agencies both compete and negotiate to set a definition of acceptable risk.

From a theatrical vantage point, social arenas of risk are populated by sundry groups of actors. Palmlund (1992) proposes the existence of six 'generic roles' in the societal evaluation of risk, each of which carries its own dramatic label: risk bearers, risk bearers' advocates, risk generators, risk researchers, risk arbiters and risk informers.

Risk bearers are victims who bear the direct costs of living and working in hazardous settings. In the past, those who are impacted most have rarely asserted themselves and have therefore remained on the margins of risk arenas. More recently, however, as can be seen in the rise of the environmental justice movement, risk bearers have become empowered and must increasingly be regarded as notable players. *Risk bearers' advocates* ascend the public stage to fight for the rights of victims. Examples include consumer organisations such as those headed by Ralph Nader and Jeremy Rifkin, health organisations, labour unions and congressional/parliamentary champions. They are depicted as protagonists or heroes. *Risk generators* – utilities, forestry companies, multinational chemical and pharmaceutical companies, etc. – are labelled as antagonists or villains since they are said by advocates to be the primary source of the risk. *Risk researchers*, notably scientists in universities, government laboratories and publicly funded agencies, are portrayed as 'helpers' attempting to gather evidence on why, how and under what circumstances an object or activity is risk-laden, who is exposed to the risk and when the risk may be regarded as 'acceptable'. On occasion, however, risk researchers have become identified with risk generators, particularly if their findings support the latter's position. *Risk arbiters* (mediators, the courts, Congress/Parliament, regulatory agencies) ideally stand off-stage seeking to determine in a neutral fashion the extent to which risk should be accepted or how it should be limited or prevented and what compensation should be given to those who have suffered harm from a situation judged to be hazardous. In reality, risk arbiters are rarely as neutral as they should be; instead, they frequently tend to side with risk generators. Finally, *risk informers*, primarily the mass media, take the role of a 'chorus' or messengers, placing issues on the public agenda and scrutinising the action.

Renn (1992) suggests a hybrid of several of these roles: the *issue amplifiers* who observe the actions on stage, communicate with the principal actors, interpret their findings and report them to the audience. Environmental 'popularisers' such as Paul Ehrlich, Barry Commoner, Jeremy Rifkin and Jonathan Porritt are prime examples of this.

Hilgartner and Bosk depict the interaction among different arenas of public discourse as characterised by several key features. First, these multiple arenas are connected by a complex set of linkages both social and organisational. As a result, activities in each arena thoroughly propagate throughout the others. Second, one finds a huge number of 'feedback loops' that either amplify or dampen the attention given problems in public arenas. Consequently, you find a relatively small number of successful social problems which occupy much of the space in most of the arenas at the same time. This synergistic pattern is typical of policy-making on matters relating to risk and the environment.

In their study of 228 Washington-based 'risk professionals', Dietz and Rycroft (1987) found a policy community with a dense network of communication which stretched across environmental groups, think-tanks, universities, law and consulting firms, corporations and trade associations, the EPA and other executive agencies. Environmental organisations were especially active in outreach activities including contacts with corporations and trade associations with whom 85 per cent of respondents communicated in a typical month. Similarly, the personnel flows across organisations, another component of the exchange network, was substantive, although working for an environmental group led to a low probability of finding employment with one of the other groups.

Dietz and Rycroft depict the environmental risk policy system as a hybrid in the sense that it has a strong base in science but at the same time is driven by the ideological conflict between environmentalists and corporate and governmental participants. This creates a measure of volatility insomuch as science is the cornerstone of the system yet many key decisions are resolvable only in political terms. Nevertheless, the picture that emerges from this survey study is one of a policy community which is permeable but nevertheless closely linked and oriented towards a shared discourse on issues relating to environmental risk. Among other things, this means that any approach to risk that attempts to emphasise sociocultural facts over physical ones will probably be considered off target and therefore inappropriate for inclusion on the shared agenda of the risk professionals (Dietz and Rycroft 1987: 114).

POWER AND THE SOCIAL CONSTRUCTION OF ENVIRONMENTAL RISK

Freudenburg and Pastor (1992) have asserted that the social constructionist approach to risk suggests closer attention to those variables that sociologists have found to be associated with the exercise of power. In a similar fashion, Clarke and Short (1993) observe that constructionist arguments – in contrast to psychology and economics – tend to focus on how power works in framing terms of debate about risk.

Both sets of authors share the belief that this relationship is especially important because official viewpoints, with their significantly greater access to the mass media, strongly suggest that public fears regarding technical risks are clearly irrational; that is, claims about public irrationality are in themselves ways of framing risk issues. By implication, policy formulations which originate with the community of risk professionals which we discussed in the previous section are presented as rational, objective assessments of what is considered safe and what is not. If this view is accepted, then the central task is said to be educating the public to realise that they are over-reacting and that the risk attached to nuclear power, herbicides, bioengineered organisms, etc., are not really the hazard that they appear to be. In order to allay public fears, risk analysts develop quantitative measures through which to compare the risks inherent in different policy choices and their relative costs and benefits (Nelkin 1989: 99).

This is not to imply that the people are always right and the knowledge of the experts invariably 'brittle' (Wynne 1992: 276). Rather, a social constructionist perspective would argue that each represents a competing frame but the dominant rationality which comes from the risk establishment is superimposed over the popular frame due to a power differential. Thus, Wynne (1992: 286) demonstrates in the case of a public controversy over the herbicide 2,4,5-T in the United Kingdom that the firsthand empirical knowledge of farm and forestry workers was directly relevant to an objective risk analysis, but that this knowledge was rejected by scientists, thus denigrating and threatening their social identity.

Nowhere is this differential more evident than at public information meetings or hearings which are routinely stage-managed by risk generators and arbiters. At the public meetings concerning the building of the sewage detention tanks described earlier in this chapter (see pages 96–7), members of the public works department, local politicians (who strongly supported the project) and representatives of the private engineering firm who had recommended the building of the tanks all sat together on the elevated stage of the auditorium whose perimeters were adorned with charts, blown-up photographs and other 'props'. We citizens were restricted to a single question with no follow-up. Those who queried the suitability of the project were alternately bullied and patronised. On contentious issues the presenters did not hesitate to introduce a ream of previously unseen statistical evidence which we had no way of confirming or denying without days or weeks of further research. Richardson *et al.* (1993) observed many of the same structural elements in the conduct of a set of environmental public hearings in 1984 on the proposed building of a bleached kraft pulp mill in Northern Alberta.[4] For example, the members of the Alpac EIA Review Board who were conducting the hearing sat at a table facing the public, sometimes on a stage. At one or several tables to the direct right of the Board were the representatives of Alberta-Pacific Forest Industries (Alpac), the company which sought to build the mill, their technical experts and their lawyer. Numerous Alpac consultants were scattered throughout the proceedings. Presenters were required to speak into microphones through which their words were recorded.

Kaminstein (1988) argues that embodied in the public presentation of scientific information at meetings concerning the health and safety aspects of toxic waste dumps is a rhetoric of containment which restricts discussion, avoids tough questions and pursues its own agenda. Drawing on three years of observation of EPA meetings held to inform residents of Pitman, New Jersey, about the steps which were being taken to clean up the Lipari landfill, the site of one of the worst dumps in the United States, Kaminstein concludes that residents were not so much informed or persuaded as controlled and defeated. The primary tool which scientific experts associated with the EPA and the Centers for Disease Control used to stifle citizen initiatives was toxic talk – talk which stifles discussion and smothers public concern. The rhetoric of containment has multiple elements.

First, as was the case with the detention tank meetings, residents were bombarded with technical information. At one meeting, EPA officials gave out documents totalling forty-four pages. Those in attendance were expected to assimilate an array of data, charts, graphs, tables and a slide show in rapid succession. At the same time, the facts that residents wanted were never available and no explanation or interpretation was given as to the information which the consultant scientists presented.

The physical setting of the meeting room was also very similar to that experienced by those attending the detention tank sessions. At the front of the room was a large dais raised about two feet, a long table and nine large, high-backed chairs on which the scientists sat, creating a physical and psychological distance from the audience. Various dramatic props, for example, an enlarged photograph of an air-monitoring vehicle which looked like a recreational camper, were employed as rhetorical devices to pacify the residents and enhance the power of those in charge of the meeting.

The factual presentation style used by EPA officials and scientists was abstract, impersonal and technical, thus creating an impression of professional neutrality. It was the activist residents who became angry and confrontative, allowing officials to dismiss them as 'emotional'. Questions which dealt with the geology and hydrology of the area, future tests and plans for the clean-up were addressed but those which dealt with health risks were avoided or deflected. Officials and scientists used language in their presentations which was technical, ambiguous and intellectual, making it impossible for any meaningful dialogue to develop between experts and residents over the nature and magnitude of the risks faced by the community of Pitman.

Toxic talk techniques such as this are strategically successful if ethically reprehensible. It allows scientific experts and government officials to direct the discussion, set the risk agenda and discourage future citizen participation. Popular concerns and risk frames are subordinated to those which are preferred by the powerful in society. As Kaminstein (1988: 10) notes, these kinds of exclusionary devices permit agencies such as the EPA to legally fulfil their mandate to hold public meetings while at the same time leaving residents feeling that they are fighting a losing battle just to be heard.

That is not to say that members of the public never attempt to assert themselves in official settings such as these. For example, in the Alberta case, some participants fought to wrest control from regulators over the scope of the review, the venues and over definitions of legitimacy, as well as attempting to subvert the dominant discourse which was imposed by the pro-development forces (Richardson *et al.* 1993: 47). However, the constraints of the hearing process normally make effective citizen participation difficult, especially since the situation is structured so as to prevent public argument and reinforce the power of institutions.

Institutional risk analysts and regulators also exercise power on a broader plane. Structurally, they control the official risk agenda, acting as gatekeepers who are well placed to determine which issues are included or excluded from public discourse. For example, in the 1980s, imbued with the deregulatory climate within the Reagan administration (which was supported by senior EPA managers), Congress fatally slashed the budget of the Office of Noise Abatement and Control (ONAC), thereby also dooming most state and local noise abatement programmes (Shapiro 1993). Despite the continued risk posed by noise pollution to human health and environmental aesthetics, the issue stalled for lack of government action and only recently showed some signs of revival. In such circumstances, the risk itself does not diminish (in the case of noise pollution it in fact increased) but the risk establishment is able to manipulate its progress on the action agenda.

Freudenburg and Pastor (1992:403) note that the social constructionist approach to technological risks would do well to look at other variables which sociologists have previously found to be associated with power. Thus gender may be significant here, insomuch as the scientific experts and bureaucratic officials who practise the rhetoric of containment are usually men while local citizen groups are disproportionately compose women, many of whom lack power and authority in public life. Similarly, members of racial and ethnic minorities are routinely dismissed and discredited by the risk establishment, an experience which has led to the recent blossoming of the environmental justice movement. The relationship between power, inequality and the social construction of risk is equally evident in communities which have been marginalised by positions of economic, geographic or social isolation (Blowers *et al.* 1991).

RISK CONSTRUCTION IN CROSS-NATIONAL PERSPECTIVE

Finally, risk construction varies cross-nationally according to a number of different factors: the organisation of political and administrative structures, historical traditions and cultural beliefs. There are several good illustrations of this to be found in the three case studies which are presented later in this book (see chapters 7–9). Consider, for example, the different fate of bST, the generically engineered bovine growth hormone, in the United States, where it has been approved for use, versus Western Europe where it has effectively been banned. Or note the contrasting patterns of action/inaction by science and government in Sweden, the United States, Canada, Britain and Germany with regard to acid rain.

Within the field of risk analysis, perhaps the best comparative study is Jasanoff's (1986) report entitled *Risk Management and Political Culture*. Drawing on case studies of national programmes for controlling carcinogens in several European countries, Canada and the United States, she concludes that cultural factors strongly influence goals and priorities in risk management. In (West) Germany, the favoured approach has been to delegate resolution of all risk-related issues to technical experts. Jasanoff does not discuss it but even where a risk subject is strongly contested, technical rationality is applied in the form of a 'technology assessment' which includes representatives from government, industry and social movements (see Bora and Dobert 1992). In Britain and Canada, risks are examined through a mixed scientific and administrative process but scientific uncertainties are not always publicly broadcast. By contrast, in the United States risk determination has a much more public face surfacing in a wide variety of administrative and scientific fora. While this can produce greater analytical rigour and more democratic and informed public participation, it can also lead to more polarisation and conflict and thus to political stalemate.

Using the comparative method suggested by Jasanoff, Harrison and Hoberg (1994) compared government regulation in Canada and the United States of seven controversial substances suspected of causing cancer in humans: the pesticides Alar and alachlor, urea-formaldehyde foam insulation, radon gas, dioxin, saccharin and asbestos. Each country's approach was weighed according to five criteria for effectiveness: stringency, and timeliness of the regulatory decision; balancing of risks and benefits by decision-makers; opportunities for public participation; and the interpretation of science in regulatory decision-making.

As had Jasanoff, the researchers found that there were two contrasting regulatory styles. In each case

> there was more open conflict over risks in the United States than Canada, with interest groups, the media, legislators and the courts playing a much more important role south of the border. The regulatory process in Canada tended to be closed, informal, and consensual, in comparison with the open, legalistic, and adversarial style of the U.S.
>
> (Harrison and Hoberg 1994: 168)

Both styles are said to have risks and benefits. The Canadian system is more conducive to scientific caution and formal democratic control but it lacks accountability, making it easier for political decisions to be cloaked in scientific arguments. The American system is more open but also more conflictual and vulnerable to interest group pressures and, as a result, less dependent upon scientific expertise.

This comparative research provides further evidence that risk determination and assessment are socially constructed. National political structures and styles can be seen to have as much to do with deciding which environmental conditions will be judged to be risky and actionable as the nature of the scientific claim itself. Consequently, fundamentally sound environmental claims may be stopped at the second or third phases of the model presented in Chapter 2, either due to collusion between regulators and scientists or because of political pressure from interest groups, either within or opposed to the environmentalist perspective.

6

NATURE, ECOLOGY AND ENVIRONMENTALISM

Constructing environmental knowledge

Along with risk, a second socially constructed product is knowledge about the environment and the problems which bedevil it. As Elizabeth Bird (1987: 260) has pointed out, our understanding of environmental problems is in itself 'a social construction that rests in a range of negotiated experiences'. Indeed, there are multiple ways of representing nature and the environment from the scientific to the mystical. Rather than a fixed entity, the environment is a fluid concept which is both culturally grounded and socially contested.

As we saw in Chapter 1, a major dichotomy is between an anthropomorphic, growth-centred view of the environment which is embedded in mainstream culture and an ecocentric, holistic interpretation which is central to 'green' thinking. Environmental definitions may also differ substantially along class and ethnic lines, varying in accordance with historical circumstances and shared life experiences (Lynch 1993: 108–9). How we construct environmental knowledge subsequently becomes the basis for contesting claims as to basic rights, responsibilities and responses towards technology, nature and society.

Environmental historians have long taken the lead in exploring ways in which the concept of 'nature' is socially contingent. As Bailes (1985: 5) puts it, environmental history deals with 'the dialectic between nature and culture, the interaction of humans with the rest of nature through time'. It is appropriate, then, to begin a discussion of the social construction of environmental knowledge by examining the concept of nature itself and how this came to be fundamentally transformed during the latter decades of the nineteenth century.

SOCIAL CONSTRUCTION OF NATURE: THE 'BACK TO NATURE' MOVEMENT IN EARLY TWENTIETH-CENTURY AMERICA

As Europe and America became increasingly urbanised at the close of the nineteenth century, views towards nature began to undergo a major transformation. In particular, the concept of 'wild nature' as a threat to human settlement which had long predominated gave way to a new, intensely romantic depiction in which the wilderness experience was celebrated.

The traditional image of nature and its inhabitants as frightening is reflected in much of our past and present 'mythical' literature. For example, wolves play a central role in fairy-tales such as *Little Red Riding Hood* and *Peter and the Wolf* and, most recently, in the Disney film version of *Beauty and the Beast*, making the woods a dangerous place for children to wander alone. Similarly, readers are advised to keep out of the forest at night to avoid spectres such as the Headless Horseman in *The Legend of Sleepy Hollow*. Civilisation is depicted here as the conversion of untamed natural landscapes into a more refined pastoral setting. Note, for example, Tolkien's contrast in *Lord of the Rings* between the gentle, civilised, rolling vistas of the hobbit settlements and the wilder, darker world of the forest and mountains inhabited by walking trees, orcs and other threatening creatures.

This unfavourable attitude towards untamed nature was especially heightened during the settlement of the American frontier:

> Wild country was the enemy. The pioneer saw as his mission the destruction of the wilderness. Protecting it for its scenic and recreational value was the last thing frontiersmen desired. The problem was too much raw nature rather than too little. Wild land had to be battled as a physical obstacle to confront and even to survive. The country had to be 'cleared' of trees. Indians had to be 'removed'; wild animals had to be exterminated. Natural pride arose from transforming wilderness into civilization, not preserving it for public enjoyment.
>
> (Nash 1977: 15–16)

By the last part of the nineteenth century, however, a revised view of unmodified nature had emerged. Rather than a threat, wilderness was now seen as a precious resource. This view was especially strong in the United States where the frontier was on the verge of closing. In the Eastern portions of the country, natural landscapes were rapidly disappearing as urban growth proceeded. Urban expansion, in turn, seemed to produce a surfeit of noise, pollution, overcrowding and social problems. In this context, unspoiled natural settings took on a special meaning; that is, the stress of city living created a rising tide of nostalgia among the urban middle classes for the joys of country life and outdoor living.

Schmitt (1990) has identified a 'back to nature' movement which flourished in the United States from the turn of the century to shortly after the First World War. This movement or 'wilderness cult' (Nash 1967) encompassed a wide range of activities including summer camps, wilderness novels, country clubs, wildlife photography, dude ranches, landscaped public parks and the Boy Scouts. While it was not the only factor, this nature-loving sentiment played a significant role in the creation of the natural parks system. In the process, wild nature was transformed from a nuisance to a sacred value. As Charles Adams wrote in the *Naturalist's Guide to the Americas* (1926), the wilderness, like the forests, was once a great hindrance to our civilisation; now, it must be maintained at great expense because society cannot do without it (Schmitt 1990: 174).

It is quite clear from Schmitt's and other accounts that this back to nature movement and the 'arcadian myth' which it promulgated was socially constructed. While its supporters had mixed motives, they generally shared a belief that a return to nature represented a more wholesome set of values from those to be found in the increasingly corrupt environment of the city. Claims about the virtue of nature were made in each of the major institutions of the day. Leading American educators such as G. Stanley Hall, Francis Parker and Clifton Hodge actively encouraged nature study in the schools as a means of counteracting urban vices and building character. Religious educators, convinced that Americans could best find Christian values out of doors, promoted a form of pastoral Christianity in a number of ways: nature sermons, outdoor church camps, sponsorship of Scout troops, and so on. Nature journalists published a steady stream of nature lore, essays, outdoor pictures and literary tales (e.g. Jack London's *Call of the Wild*), all celebrating the lure of wild nature. The case for wilderness preservation was taken on by a clutch of new conservation organisations from the Sierra Club (1892) to the Wilderness Society (1935). This preservationist sentiment was especially strong among bird-watchers and ornithologists who participated in a series of crusades for over fifty years in both Britain and the US to protect wild birds from hunters, poachers, feather merchants and other enemies (see Doughty 1975).

Despite widespread popular acceptance, many of the initiatives undertaken by preservationists and other nature promoters were vigorously contested. Nowhere is this more evident than in the Hetch–Hetchy controversy of 1901–14 when resource conservationists as represented by US Chief Forester Gifford Pinchot triumphed over preservationists represented by John Muir, gaining official approval for a plan to create a reservoir in a valley deep in Yosemite Park to supply water and power to the city of San Francisco. This conflict, in fact, prefigures a number of the environmental clashes of the present era.

The social construction of the back to nature movement incorporated each of the six factors which I have deemed to be necessary for the accession of a specific environmental problem on to the public policy agenda. Although it was not primarily constructed on a scientific foundation, many aspects of the movement were given scientific legitimation.

Efforts to save the redwood forests in the first part of this century arose initially from a desire to preserve artefacts critical to the emerging disciplines of palaeontology and palaeobotany. Especially important to this endeavour were two University of California at Berkeley professors, John Merriam and Ralph Chaney, who, in addition to pioneering the development of vertebrate palaeontology on the Pacific Coast, were key figures in the Save the Redwoods League. Preserving the redwood and sequoia forests was vital to these scientists because these twenty million-year-old species were seen as having unique evolutionary significance: a living key to unlocking the geologic history of the West Coast (Schrepfer 1983).

In the United States, the original impetus for bird protection came from a committee of the American Ornithologists Union (AOU). While this group was considered primarily humanitarian rather than scientific in its work (Doughty 1975: 103), it nevertheless linked a research interest in birds with an activist stance. Under the guidance of William Dutcher, treasurer of the AOU from 1887 to 1903, and Frank Chapman, AOU bird protection committee member and ornithologist at the American Museum of Natural History, this group played a major role in cooperating with the Audubon Society and with humane societies in informing and educating politicians and the public about the rising threat to plume birds.

Nature education was justified by using the 'recapitulation' theory of genetic psychology. This was a developmental model which claimed that the human psyche needed to recapitulate the cultural epochs of human history starting with the ancestral experiences of hunting and fishing, climbing and swimming and exploring and child's play was seen as an important part of these activities. However, it was alleged that an urban upbringing denied children the opportunity to recapitulate these prehistoric activities, thus denying them their natural maturity. As a result, children turned to gang membership and juvenile crime. Recapitulation theory eventually lost legitimacy but at the turn of the century it was seen as a valid approach which significantly influenced the applied field of playground psychology (Schmitt 1990: 78–80).

The back to nature movement gained a number of prominent political and institutional sponsors. None was more important than Teddy Roosevelt who, as Governor of New York and then as President, became a staunch advocate of wildlife preservation. Roosevelt's philosophy of conservation may often have differed from that of John Muir but there was no doubt that his patronage was crucial for the success of many campaigns to protect nature. Another key supporter was David Starr Jordan, the first president of Stanford University, whose voice in support of nature study gave the movement credibility and prestige (Lutts 1990: 28). A number of important figures in the movement were based in public institutions: the American Museum of Natural History, the Smithsonian, the Carnegie Institution and the New York Zoological Society to which they were able to bring considerable resources – money, publicity, prestige to their preservationist and other activities on behalf of nature.

It was from these institutions also that many of the key popularisers of nature protection originated. For example, the movement to save the redwoods contained several leading scientific popularisers of the day: Madison Grant,[1] a New York lawyer and author; Edward E. Ayer, head of the Chicago Museum of Indian History; Gilbert Grosvenor, founder of the National Geographic Society; and Fairfield Osborn, a key figure in the growth of the New York Museum of Natural History (Schrepfer 1983: 41). Perhaps the highest profile populariser (next to Teddy Roosevelt) was William T. Hornaday, for many years director of the New York Zoological Society (Bronx Zoo) who was a major force in lobbying Congress to tighten hunting regulations. Hornaday, a tireless self-promoter, wrote several widely distributed volumes on wildlife preservation as well as numerous articles in the *New York Times* and other popular publications. John Muir, the founder of the Sierra Club, was a charismatic promoter of wilderness protection who waged the country's first nationwide environmental publicity campaign during the Hetch–Hetchy controversy.

Popularisers such as Hornaday and Muir, as well as other claims-makers within the broad back to nature movement, were highly successful in garnering media attention. In this age of magazines, nature study essays and outdoor adventures were frequently featured in *Outlook, The Atlantic Monthly, Forest and Stream, Saturday Review, National Geographic* and other popular periodicals. In addition, various campaigns initiated their own publications, some of which developed a large readership. For example, the bird preservation movement spawned *Bird Lore*, the *Audubon Magazine* and other similar periodicals. *Boy's Life*, a monthly picture magazine which capitalised on the growing popularity of scouting, sold a cumulative total of forty-one million issues from 1916 to 1937 (Schmitt 1990: 111). One environmental campaign, the crusade to save Niagara Falls (1906 to 1910), was waged primarily in the pages of American popular magazines, notably the *Ladies Home Journal*; it resulted in over 6,500 letters written in support of the preservation of the Falls (Cylke 1993: 22).

The back to nature movement drew upon a deep wellspring of existing cultural sentiments and in turn created a number of readily identifiable symbols and icons: the horse, Black Beauty,[2] the California redwood trees, the Grand Canyon, Old Faithful geyser in Yellowstone National Park and even Smokey the Bear. Some of these were real, others fictional creations. None the less, as Schmitt (1990: 175) notes, 'those who dealt in symbols and myths found the wilderness a major force in shaping American character'.

Finally, while the back to nature movement was not entirely driven by economic motives, there is nevertheless ample evidence of economic incentives which encouraged positive action on matters of nature protection, preservation, education and celebration. Support by the Southern Pacific Railroad significantly hastened legislation to establish Big Basin Redwood State Park just south of San Francisco, the first redwood forest to gain official protection. This support was extended on the premise that the railroad would be able to carry paying passengers into the park (Schrepfer 1983: 11–12). In fact, the railroads soon became prime promoters of the newly established national parks; for example, in 1919, a special information bureau of the United States Railroad Administration distributed two-and-a-half million pieces of literature extolling the natural scenic wonders of America's national parks (Schmitt 1990: 160). Stephen Mather, a wealthy borax merchant with strong ties to the railroads who became the first Director of the National Park Service in 1916, argued that tourism could be successfully integrated into a broader appeal of experiencing nature. In order to stimulate park travel, Mather promoted and even personally financed road construction in national parks such as Yosemite; in return, the automotive industry widely advertised the parks through billboards and other media (Gottlieb 1993: 32).

As Carolyn Merchant (1987) has argued, nature should not be viewed as 'some ultimate truth that was gradually discovered through the scientific processes of observation, experimentation and mathematics' but rather it constitutes a social construction which changes over time. In the first quarter of this century, the reigning ideology which emphasised nature's role as a producer of resources for developing capitalist markets was challenged by a new paradigm which assigned a positive aesthetic value to nature. While this newly emergent construction ultimately failed to fully dislodge the existing economic view, it nevertheless left behind a number of residues: bird sanctuaries, national parks, wilderness trails, landscaped cemeteries, conservation organisations and, above all, an ethic which stated that nature had more than simply a functional value.

ECOLOGY AS A SOCIAL CONSTRUCTION

A second major example of the social construction of environmental knowledge can be found in the framing and successful incorporation during the 1970s into popular discourse of the concept of 'ecology'.

Ecology has a long history prior to its ascendancy as the cornerstone of the contemporary environmental movement. Worster (1977: xiv) observes that while the term ecology did not appear until the latter part of the nineteenth century, and it took almost another hundred years for it to become a household word, the idea of ecology is much older than the name. None the less, the term was officially coined in 1866 under the name *oecologie* by Ernest Haeckel, the leading German disciple of Darwin. By ecology Haeckel meant 'the science of relations between organisms and their environments'.

The full development of plant ecology owed more, however, to plant geographers, most notably the Danish scholar Eugenius Warming who published his classic work *Plantsomfund* (The Oecology of Plants) in 1895. Warming's central thesis was that plants and animals in natural settings such as a heath or a hardwood forest form one linked and interwoven community in which change at one point will bring in its wake far-reaching changes at other points (Worster 1977: 199). This is, of course, a central message in the contemporary ecological outlook.

Bramwell (1989: 4) has hypothesised that two strands of ecology emerged from this period. One was an anti-mechanistic, holistic approach to biology which derived from Haeckel and the plant geographers; the other a new approach to energy economics which focused on scarce and non-renewable resources. Bramwell argues that when these two strands finally fused together in the 1970s, the modern age of ecology was born.

If Bramwell is correct, why did this fusion not take place earlier, notably as part of the back to nature movement at the beginning of the twentieth century?

One answer to this is provided by environmental historian Susan Schrepfer (1983), who demonstrates that throughout this period most natural scientists were blinded to the hardcore implications of ecological thinking because of a commitment to various theories of directed evolution, notably that of 'orthogenesis'. According to this paradigm, genetic change was neither random nor was it influenced to any great extent by the surrounding environment. Instead, it followed an orderly progress. It was not known what constituted the prime force behind this orthogenesis or 'straight-line evolution'; one popular explanation was that it was possibly hormonal, another that it was part of a 'cosmic design'.

Most of the leading scientific entrepreneurs of the back to nature movement – Henry Fairfield Osborn, John Merriam, Joseph Le Conte – believed in this directional evolution. As Schrepfer cautions, the scientists who led the wilderness movement from the 1890s through the 1930s rejected much of the content of social Darwinism in favour of a reform Darwinism which taught that human reasoning power liberated us from the survival of the fittest. Instead, humans were thought to have the power to actively engineer progress; for example, by fighting to save the wilderness. At the same time, humans were regarded as the highest product of directed evolution – an achievement made possible through technological innovation. It is not difficult to see how this assumption led to a fundamental optimism regarding science and technology and a reluctance to seriously question the orderly march of industrial progress.

Accordingly, it was unlikely that ecology would have any strong appeal to the preservationists who were at the scientific centre of the back to nature movement. Not only did they have an unwavering faith that technology would overcome any problem of finite resources but they regarded humans' ability to cope as irrevocably cast within the evolutionary design of nature itself.

Nevertheless, by the 1920s biological ecology was coming into its own. Two of the major figures in its development were Frederic Clements and Arthur Tansley who developed a distinctively twentieth-century branch of biology called 'dynamic plant ecology' or 'ecosystem ecology'.

Clements, a Nebraska scientist who spent most of his career as a research associate at the Carnegie Institution of Washington, is best known for his study of ecological succession. He visualised the process of succession as going from an embryonic ecological community to a more or less permanent 'climax community' which was in equilibrium with its physical environment. Once formed, it was difficult for potential plant invaders to compete successfully with established species within this climax community. However, a number of external environmental factors – forest fires, logging, erosion – might damage or destroy the climax and force succession to begin again (Hagen 1992: 27).

Arthur Tansley, a British plant ecologist, is generally credited with coining the term 'ecosystem' in the mid-1930s. Tansley strongly opposed

Clements's use of the word 'community' to describe the relationship of plants and animals within a certain locale, maintaining that it was misleading because it wrongly suggested the existence of a social order (Worster 1977: 301). Instead, he came up with the concept of the 'ecosystem' which he described in terms of an exchange of energy and nutrients within a natural system. Catton calls the ecosystem the most central and incisive concept in the foundation of modern ecology, especially in Tansley's original understanding of the term which was meant to 'unify our perceptions of nature's units' (1994: 81).

Tansley was eclectic in his interests and friendships, having, among other things, helped the social philosopher Herbert Spencer revise his *Principles of Biology* and pursued an interest in psychoanalysis by studying briefly under Freud and writing a popular book on Freudian psychology (Hagen 1992:80). He was also an entrepreneurial scientific leader who played an instrumental role in establishing the British Ecological Society in 1913 and served for twenty years as editor of the Society's *Journal of Ecology*.

McIntosh (1985) has depicted the views of Clements, Tansley and other scientific ecologists of this era as being somewhat ambivalent with regard to human society. On the one hand, there was an acknowledgement that ecology had much to contribute to the understanding of human affairs. Clements (1905: 16) observed that sociology is 'the ecology of a particular species of animal and has, in consequence, a similar close association with plant ecology'. Tansley (1939), in his second presidential address to the British Ecological Society, anticipated the establishment of a worldwide ecosystem 'deriving from interdependence' and stated that human communities 'can only be intelligently studied in their proper environmental setting'. While it is probably an exaggeration to state as did some that ecology was the scientific arm of the conservation movement (McIntosh 1985: 297–9), nevertheless many ecologists were individually active in conservation causes. Tansley himself contributed towards the campaign to establish nature reserves and later (1949) served as the first chair of the British Nature Conservancy. In the 1940s, he led efforts (mostly unfulfilled) among ecologists to establish research linkages with the four British forestry societies on the grounds that post-war plans for giant new forest plantations would cause soil fertility to suffer as well as introducing an alien feature into the aesthetics of the countryside (Bocking 1993: 92–3).

Yet at the same time, ecologists and their societies were somewhat nervous of becoming too involved in political or social issues, fearing that their scientific credibility would be damaged. Both the British and American ecological societies were reluctant to engage in overt advocacy of particular positions or in political lobbying (McIntosh 1985: 308). Any synthesis of animal and plant ecology with human ecology was discouraged by the failure of the Chicago School in the 1920s and 1930s to adequately conceptualise the field.[3]

By the early 1970s, ecology had become the theoretical cornerstone of the new and rapidly diffusing concern with the environment. Ecologists increasingly began to step outside their role as scientists to become major contributors to the environmental debate. A plethora of new terms were added to the English language; for example, ecopolitics, ecocatastrophe, ecoawareness (Worster 1977: 341). A British magazine *The Ecologist* became a centre of gravity for left-wing environmentalism under the guidance of Edward Goldsmith. Ironically, all this occurred in the midst of a deep intellectual schism within the field of biological ecology between ecosystem ecology[4] and evolutionary ecology over the concept of 'group selection', a form of self-regulation which naturally checks population growth.

There are several key factors which explain the centrality of ecosystem ecology in the rise of environmentalism in the 1970s.

First, the language and logic of ecology was linked to rising concerns about radioactive fallout, pesticide poisoning, overpopulation, urban smog and the like to produce what appeared to be an inclusive scientific theory of environmental problems. Rubin (1994) has argued that the instrumental force in effecting this transformation was a small group of influential writers and thinkers – Rachel Carson, Barry Commoner, Paul Ehrlich, Garrett Hardin – who functioned as scientific popularisers. Carson, in her book *Silent Spring*, brought the concepts of ecology, food chains, the 'web of life' and the 'balance of nature' into the popular vocabulary for the first time. Using ecology as the explanatory linchpin, she simplified a variety of problematic human/nature relationships into one 'environmental crisis' (Rubin 1994: 45). Commoner (1971) systematised Carson's observations with his four laws of ecology: 'everything is connected to everything else'; 'everything must go somewhere'; 'nature knows best'; 'there is no such thing as a free lunch'. These laws may have over-simplified ecosystem ecology but they had enormous rhetorical power. Similarly, Hardin's (1978) metaphor of the 'tragedy of the commons' found a broad audience both within the academic world and outside.

Second, the fusion of ecology and ethics first achieved in Aldo Leopold's 'land ethic' was featured prominently. The land ethic was first proclaimed in his book *A Sand County Almanac*, published posthumously in 1949. It extended ethical rights to the natural world which he regarded as a community rather than a commodity. In the 1950s, Leopold's work had a small but committed following in conservation circles but only became widely known after it was reprinted in 1968. Whereas the ecosystem had previously been largely a theoretical construct, albeit a dynamic one, now it was inculcated with moral significance. Human interference in biotic communities not only had particular effects, for example, forcing a new round of succession as Clements had suggested, but it was also defined as the wrong thing to do. This insight became especially significant with the rise of 'deep ecology' in the 1980s.

Finally, as Macdonald (1991:89) has observed, by coopting scientific ecology the environmental movement added considerably to its political strength for two reasons. First, despite the fact that ecosystem ecology was considered to be a somewhat 'soft' discipline within the natural sciences, it nevertheless allowed environmentalists to claim the authority of science for their campaigns. Second, because of its holistic perspective, ecology attracted a variety of 'seekers' such as devotees of expanded consciousness, Zen and organic food who might otherwise have had little interest in green causes. Combined with scientific ecologists, these newcomers created a potent political mix. In recent years, this alliance has been at best an uneasy one but in the early 1970s it brought the idea of an 'ecological threat' into the pervasive currents of alternative popular culture where journalists constantly troll in their search for the emergence of new trends.

Ecology, then, was transformed from a scientific model for understanding plant and animal communities to a kind of 'organisational weapon'[5] which could be used to systematise, expand and morally reinvigorate the environmental message. In the process, it acquired a new texture: more political, more universal and more 'subversive' (Sears 1964; Shepard 1969). While some scientific ecologists reacted negatively to this reconstitution of the concept, others supported it, arguing that the 'environmental crisis' demanded a new sense of social activism on the part of biological researchers. The latter became influential claims-makers, presenting a politicised vision in which the boundaries between nature and society were deliberately blurred.

Most recently, the meaning of ecology has once again undergone yet another reconstruction. Grassroots activists such as those found in the Chipko Movement in India have proposed a new alternative ecological perspective in which insight into ecosystem interrelationships is achieved by means of folk knowledge rather than scientific observation. This alternative knowledge system provides citizens' movements with 'the epistemological tools for the reconstruction of neopositivist science and for an alternative approach to the management of global ecological independence' (Breyman 1993: 137). In this context, ecology becomes part mythology, part popular science; a rallying point for opposition to the kind

119

of environmental diplomacy which predominated at the Rio Conference. As such, it represents a fresh social interpretation of a 130-year-old concept even if it is one which might be unrecognisable to Haeckel, Warming and other pioneers of scientific ecology.

RECONSTITUTING ENVIRONMENTALISM: ENVIRONMENTAL JUSTICE AS A DEFINING CONCEPT

Whereas the concept of ecology was utilised in the 1970s to join together rising concerns about toxic pollution with an ethical concern for nature, environmentalism in the 1990s has been undergoing another transformation in which the central underlying force is 'environmental justice'. This shift is occurring primarily at the grassroots level both domestically and in the Third World. While some key figures in this movement have wanted to throw off the environmental label entirely,[6] others have framed their claims to justice and equity within the context of an 'alternative' environmental movement. Environmental justice activists have not totally abandoned the legacy of the previous two decades: Commoner's industrial–ecological critique, for instance, has been one theoretical referent for this alternative explanation of the roots of the environmental crisis. At the same time, concerns about resource conservation, wilderness preservation and pollution abatement are de-emphasised in favour of issues such as the uneven distribution of resources and development and the occupational safety of minority workers.

Two solitudes: environmental vs. social justice issues

In his introduction to the recent special issue of the journal *Qualitative Sociology* on the topic of social equity and environmental activism, Alan Schnaiberg (1993: 203) rues the failure of environmental sociology to consider social inequality. As early as 1973, Schnaiberg claims, he was writing about the political necessity of incorporating elements of social justice into any proposal for environmental action but that this message fell on deaf ears. This may in part reflect shortcomings within the field as Schnaiberg suggests, but it is also a reflection of what was happening within the environmental movement itself.

In both the United States and Britain, the environmental movement has been dominated by a relatively narrow set of concerns; for example, rural planning and wildlife preservation. These are said to reflect the white, middle-class membership of the main environmental organisations. Taylor

(1993: 265) describes the alienation that people from black and ethnic minority groups in Britain have felt towards the countryside and the environmental movement, citing Agyeman's (1990: 3) observation that they have been deeply influenced by the experience of never seeing people of colour in tourist and other publications such as the *Countryside Commission News*. Similarly, ethnic minority groups have not been well known within the environmental field. In his examination of the toxic waste problem in Britain (and Ireland), Allen (1992: 220–1) observes that the social and racial mix of the country is not by any means reflected in the membership of the green movement which, until the late 1980s, seemed content to leave opposition to toxic waste disposal up to local groups.

This has been exacerbated by the tendency of the British government to employ a style of consensus politics in which a few selected environmental groups are invited to participate in policy-making in an inner circle, leaving the rest on the outside (Taylor 1993: 278–80). Even Greenpeace, Allen (1992: 223) charges, was 'seduced by the establishment fairly quickly' and now 'displays all the trappings of a multinational company or a civil service department'.

In the United States a number of health-related environmental inequities were exposed during the 1960s and 1970s but they rarely made it into the larger movement agenda. Gottlieb (1993) highlights the differential treatment given to three issues during this period: pesticide poisoning; the toxicity of lead; and uranium hazards.

For migrant farmworkers in California the explosion of pesticide use through the 1950s and 1960s created a number of health-related problems. In its successful campaigns for farmworker rights during these years, the United Farm Workers (UFW) under the leadership of the charismatic Cesar Chavez aggressively attempted to pursue the pesticide issue, initiating court action to obtain information about the chemical ingredients and to ban specific pesticides including DDT, and pressing to have pesticide-related health and safety language incorporated into UFW–grower contracts.[7] Aside from some limited assistance from the Environmental Defense Fund, the mainstream environmental movement generally avoided the question of human exposure to pesticides, focusing primarily on the impact of pesticides on wildlife, as did Rachel Carson.

During the 1960s, childhood exposure to lead paint became a significant local issue in a number of inner city communities in the United States. By 1970, Gottlieb (1993:247) notes, dozens of inner city-based community groups and coalitions were organising to address lead paint issues, primarily in East Coast cities such as Rochester, Washington, New York and Baltimore. Aided by New Left-inspired groups such as the Medical

Committee for Human Rights, the lead paint movement achieved significant visibility both locally and nationally. At this point, however, the emphasis shifted to lead levels in the air, especially as a result of the emission of leaded gas (petrol). Mainstream environmental groups such as the Natural Resources Defense Council and the Environmental Defense Fund which had previously avoided involvement in this issue put a priority on reforming the Clean Air Act, eventually forcing a ban on the sale of leaded gas. The lead paint issue did not re-emerge until the late 1980s and by then the primary claims-makers were from alternative environmental groups within the social justice movement.

Starting in the 1950s, uranium poisoning began to affect thousands of transient mine and mill workers, prospectors and residents of communities living downwind of the uranium mines. This 'radioactive colonisation' (Churchill and La Duke 1985) was concentrated among native American workers in New Mexico and Arizona. For example, a 1979 spill of radioactive tailings into the Rio Puerco in Northern New Mexico contaminated significant stretches of Navajo Indian lands. As Gottlieb (1993: 251) observes, the Rio Puerto spill occurred just weeks after the Three Mile Island accident, yet it received limited attention from policy-makers and mainstream environmentalists. Indeed, during the 1950s and 1960s, conservation groups ignored uranium issues altogether because they were perceived as occurring far from scenic wilderness sites. During the following decade, environmental groups were primarily concerned with nuclear power as an alternative energy choice, although the anti-nuclear movement had begun to organise. Only recently have some groups accorded the toxic threat to Indian lands a higher priority.[8]

In each of these three cases, mainstream environmental groups focused on separate though often parallel concerns defining them in 'environmental' rather than 'social justice' terms (Gottlieb 1993: 253). In constructionist language, they established 'ownership' of the problems on behalf of a primarily upper-middle-class or elite Anglo constituency. On a more general level, they focused mainly on regulation or containment rather than seeking to subvert the social order in order to bring about a form of social reconstruction which would benefit 'have not' constituencies (Hofrichter 1993: 7).

In what proved to be somewhat of an anomaly, the Conservation Foundation, an organisation whose brief focused largely on research and education, convened a conference in Woodstock, Illinois, in November 1972 to explore the themes of race, social justice and environmental quality. At this gathering, urban planner Peter Marcuse, son of the famed social philosopher Herbert Marcuse, presciently warned participants that divorcing equity and social justice concerns from the environmental agenda threatened to create a permanent rupture (Gottlieb 1993: 253). It would be another decade-and-a-half, however, until this rupture started to reach the public eye.

Contestation: the rise of the environmental justice movement

In the United States, the environmental justice movement did not emerge until the early 1980s. As Bullard (1990: 35) notes, this new-found activism 'did not materialize out of thin air nor was it an overnight phenemenon'. Rather, it was the result of a growing hostility by urban blacks in the US to the siting of toxic landfills, garbage incinerators and the like in neighbourhoods or communities with predominantly minority populations. In the 1970s this mobilisation was confined largely to the local context but within the decade it spread to a wider theatre as the struggle for environmental equity was presented as a fight against 'environmental racism'.[9]

There are several key milestones in the emergence and growth of the environmental justice movement during this period.

In 1987, the Commission for Racial Justice of the United Church of Christ (UCC) issued an influential report entitled *Toxic Wastes and Race in the United States* which documented and quantified the prevalence of environmental racism. The UCC report firmly established the 'grounds' for this claim by setting out the magnitude of the problem in numerical terms. Among its findings was the revelation that three out of five black Americans live in communities with uncontrolled toxic waste sites. Furthermore, blacks were heavily over-represented in those metropolitan areas with the greatest number of such sites: Memphis, Tennessee; St Louis, Missouri; Houston, Texas; Cleveland, Ohio; and Chicago, Illinois, with over a hundred each. Hispanics, Asian Americans and native peoples were similarly over-represented in high-risk communities. This confirmed a study conducted four years earlier by the US General Accounting Office which reported that three of the four largest commercial landfills in the South were located in communities of colour.[10]

Also crucial in establishing the dimensions of environmental racism was the research of sociologist Robert Bullard. In 1979, Bullard, then teaching at the predominantly black Texas Southern University in Houston, was asked by his wife, a lawyer, to conduct a study on the spatial location of all of the municipal landfills in that city in order to provide data for a class action lawsuit which she was arguing. Bullard confirmed that toxic waste facilities, not only in Houston but elsewhere in America, are most likely to be found in black and Hispanic urban communities. In a series of journal

123

articles beginning in 1983 (many co-authored with Beverly Wright) and in his book, *Dumping in Dixie* (1990), Bullard documented these environmental disparities and the mobilisation of the environmental equity movement. Even more than was the case for the UCC report, Bullard's work established the 'warrants' for this problem, arguing that action was justified in order to reclaim for minorities 'the basic right of all Americans – the right to live and work in a healthy environment' (1990: 43). Bullard has since become a key leadership figure as indicated in 1992, when he was chosen by the Clinton administration to participate in the presidential transition process as a representative of the environmental justice movement (Miller 1993: 132).

In January 1990, Bunyan Bryant and Paul Mohai, professors in the School of Natural Resources at the University of Michigan, organised the University of Michigan Conference on Race and the Incidence of Environmental Hazards with papers from twelve scholar-activists. Among the follow-up strategy of the Michigan Conference was a series of meetings in Washington with key government officials including William Reilly, Adminstrator of the EPA, and Congressman John Lewis (Bryant and Mohai 1992).

Third, under the sponsorship[11] of the UCC Commission for Racial Justice, the First National People of Color Environmental Leadership Summit was held in October 1991 in Washington, DC. At this summit, three strands of environmental equity were identified (Lee 1992): *procedural equity* (governing rules, regulations and evaluation criteria to be applied uniformly), *geographic equity* (some neighbourhoods, communities and regions are disproportionately burdened by hazardous waste) and *social equity* (race, class and other cultural factors must be recognised in environmental decision-making).

Gottlieb (1993: 269) credits the summit with advancing the environmental justice movement past a 'critical threshold in definition' both by ratifying a common set of principles and by identifying a new kind of environmental politics of inclusion.

Organisationally, the movement has been held together in a number of decentralised, loosely linked communication networks of umbrella groups, newsletters and conferences (Higgins 1993: 292) rather than the top-down, professionalised configuration typical of mainstream environmentalism. This has its roots in the formation in the early 1980s of several network-based 'anti-toxics' groups – the Citizens' Clearinghouse for Hazardous Wastes and the National Toxics Campaign. More recently, the emphasis has shifted from national to regional grassroots networks, as epitomised by the Southwest Network for Environmental and Economic Justice.

These new minority grassroots groups work closely with umbrella organisations such as the above-mentioned anti-toxics clearinghouses but they usually keep their distance from established national environmental groups. In part, this reflects the perception that these national groups have not helped them in the past and are reluctant both to offer minorities real input in decision-making and the chance to attain middle- and upper-level management positions (Taylor 1992).

In Britain, minority entry into the environmental arena has taken a somewhat different form. While most of the minority environmental groups are grassroots operations with strong links to churches, community centres and other social justice organisations, they have not, for the most part, arisen from specific struggles against toxic siting,[12] as has been the case in the United States.

Perhaps the most prominent minority environmental group on an empirical level is the Black Environmental Network (BEN) founded in September 1988 at a conference sponsored by Friends of the Earth and the London Wildlife Trust. Its main goal has been to secure funds to help minority communities undertake environmental projects, ensure access to the countryside and to help mainstream environmentalists become more sensitive to minority concerns (Taylor 1993: 271).

While minority environmental activists in BEN and other groups have incorporated the principle of environmental racism into their discourse, it coexists somewhat uneasily with other aspects of environmentalism and community action. Taylor (1993) depicts these organisations as under-mobilised and overly dependent on funding from corporate and environmental sponsors; for example, in 1991, BEN directed its corporate funding towards a symbolic tree-planting project. Neither BEN nor the Karibbean Ecology Trust, the other major minority environmental organisation in Britain, have dues-paying members, a significant contrast to the American case where a vigorous grassroots membership has been a central feature of the movement.

In recent years, the environmental justice movement has expanded its charter to incorporate the exploitation of Third World peoples. Much of the interaction between grassroots activists from the United States and their counterparts in the South has taken place in the context of the United Nations; for example, at the Rio Summit and its preparatory meetings. Environmental justice activists from the US also participated in a 1992 meeting hosted by the Third World Network in Malaysia which focused on toxic waste. These networking activities with Third World activists have somewhat ironically moved environmental justice leaders back on the path to a renewed ecological awareness. Thus Vernice Miller, a co-founder of the group West Harlem Environmental Action, now describes environmental justice as a 'global movement that seeks to preserve and protect global ecosystems' (1993: 134). It will be interesting to see whether this accommodation of global ecology with environmental justice does indeed represent an 'evolution' of the movement's thinking as Miller suggests or whether it will in fact direct the movement away from its roots in community activism, thus provoking a future internal conflict.

Environmental justice is a socially constructed frame which lays out a set of claims concerning toxic contamination in terms of the 'civil rights' of those affected rather than in terms of the 'rights of nature' (Nash 1989). Capek (1993) identifies four major components of this environmental justice frame: the right to obtain information about one's situation; the right to a serious hearing when contamination claims are raised; the right to compensation from those who have polluted a particular neighbourhood; and the right of democratic participation in deciding the future of the contaminated community. Each of these components represents a specific claim which has been rhetorically formatted in the language of 'entitlement' (Ibarra and Kitsuse 1993). As noted above, these are by no means the only interpretive frames that are possible: for example, The First National People of Color Environmental Leadership Summit has proposed a set of global entitlements which, among other things, claims the right to political, economic, cultural and environmental self-determination for all peoples.

CONCLUSION

In this chapter, I have demonstrated that three key concepts – nature, ecology and environmentalism – are by no means fixed in meaning, but instead are both socially constructed and contested. As Alexander Wilson (1992: 13) has observed about the first of these, 'nature ... is not a timeless essence, but, rather, humans and nature continually construct one another'.

As it happens, the history of the environmental movement, especially in America, falls into three phases which roughly correspond to the reworking of the meaning of each of these concepts. In its original manifestation as a back to nature movement at the turn of the century, environmentalism was primarily concerned with transforming the idea of 'wild nature' into a new, romantic, spiritually charged construct in which the wilderness experience was duly celebrated. In the wake of the phenomenal public impact of Rachel Carson's book *Silent Spring* in the 1960s, the concept of ecology was appropriated from its niche within applied biology, morally invigorated, and made the explanatory linchpin in the battle against pollution. More recently, the central identity of environmentalism has been reformulated by an increasingly influential and vocal rump of grassroots activists such that the rights of nature have been supplanted by the rights of minorities to environmental justice.

At the same time, there have been many other notable attempts to redefine environmentalism to reflect some other set of unique experiences. For example, gender has been utilised as the basis of one such reformulation, notably in the form of an 'ecofeminist' critique. Emerging in the late 1970s as an attempt to join radical feminist and ecological concerns, ecofeminism has reinterpreted psychology (Griffin 1978), the history of science (Merchant 1980) and religion (Spretnak 1982) from the vantage point of a 'socially constructed connection between women and nature' (Epstein 1993: 148).

Environmental knowledge, then, is highly contingent and open to constant revision in response to changing cultural currents. As Wilson (1992: 87) notes, 'the culture of nature – the ways we think, teach, talk about and construct the natural world – is as important a terrain for struggle as the land itself.'

7

ACID RAIN

From scientific curiosity to public controversy

One of the most technically complex and politically charged environmental controversies in recent times is that which focuses on the damage by and control of 'acid rain'. Acid rain is formed when sulphur dioxide and nitrogen oxide gases, primarily from coal-burning electrical power plants, industrial smelters and automobile emissions, combine with moisture from the atmosphere to form powerful acids. When these acids return to the earth as rain, snow or other forms of precipitation, or in dry form, they are thought to provoke a battery of damaging environmental effects in regions where the soil, rocks and lakes are poor buffering or neutralising agents. The damage caused by acid rain, both directly and indirectly, is said to include the acidification of lakes and streams resulting in the death of fish and other freshwater marine life, a severe decline in the growth of high altitude forests, the disruption of microbial ecosystems and the crumbling of statues, monuments and other historical artefacts.

Over the past two decades, acid rain has been transformed from 'an esoteric topic of scientific research in certain specialised fields of ecology and atmospheric chemistry into a household word' (Regens and Rycroft 1988:3). In this respect, it parallels the transformation of little-known medical scientific conditions into fully developed social problems.

As an environmental problem, acid rain has several distinguishing characteristics. First, it is a global problem with widely dispersed multiple impacts (Ellis 1989). This means that it occurs in a number of geographic locations rather than concentrating at a single impact site. Second, it is a transboundary environmental problem originating in one political jurisdiction and impacting another. This has ensured that any attempts to regulate acid emissions have been seriously contested, since the pollutant does not directly share in the benefits of reduction. Third, in contrast to the toxic waste at Love Canal or the nuclear accidents at Chernobyl and Three Mile Island, the substances which cause acid rain are found throughout nature, although they are released into the air in quantities which upset the balance of nature (Gould 1985: 4). Opponents of acid rain regulation have frequently seized upon this point, arguing that natural factors such as volcano eruptions, rotting vegetation and forest fires are more culpable than fossil fuel emissions.

ASSEMBLING THE CLAIM

While the perception that acid rain represents a serious environmental problem has its roots in science, scientific discovery alone was not sufficient to assemble a viable claim.

As far back as 1852, Robert Angus Smith, a British chemist, coined the term acid rain to describe how the high acidity levels of rain were explained by the presence of sulphuric acid which he attributed to emissions of soot from coal-burning sources in Liverpool, Glasgow and other British industrial centres. Smith speculated that a wide variety of effects were produced by this acid rain from serious damage to trees and crops and the corrosion of metals to a rise in the human mortality rate in Glasgow. Even though he rose to the position of Britain's first 'Alkali Inspector', i.e. chief of air pollution control, Smith was unable to translate his work into any concentrated political action. In the years following the publication of Smith's book *Air and Rain* (1872), a surprisingly substantial amount of work was carried out in Europe on different symptoms or problems associated with acid precipitation but none of the researchers was able to integrate findings from the multiple scientific field involved or recognise the complete problem of acid rain as it is understood today (Kowalok 1993: 14–15).

When scientific concern over acid rain finally coalesced a century later, it originated in three seemingly unrelated areas of research: limnology, agricultural science and atmospheric chemistry (Cowling 1982).

In the 1950s, Eville Gorham, a Canadian ecologist at Dalhousie University who had come across Smith's work, documented the effects of acid precipitation, first in the English Lake District and then back in Canada in the area adjacent to the giant Inco nickel mining smelter in Sudbury (Gorham 1955). However, Gorham's published work focused on unfamiliar aquatic plants and appeared in highly specialised scientific journals. It ended prematurely in 1962 when Gorham moved to the University of Minnesota. This lack of recognition, Boyle and Boyle (1983:34) insist, resulted in at least a two-year lag in both scientific and public awareness of acid precipitation and its ecological significance.

In 1963, an interdisciplinary American scientific team led by Gene Likens and Herbert Bormann initiated a study of a small watershed within an experimental forest area in New Hampshire. Despite the remoteness of the location from pollution sources, the researchers found compelling evidence of highly acidic rains. This research was not, however, widely reported until a decade later (Likens *et al.* 1972).

Of more immediate impact was the work of Svante Oden, a Swedish soil scientist. Oden, now widely regarded as the 'father of acid rain studies' (Park 1987: 6), not only found that the acidity levels of precipitation were increasing in Scandinavia but he was the first to definitively link source and receptor areas. He was able specifically to demonstrate that much of the acid rain in Scandinavia was not of local origin but rather originated with sulphur dioxide emissions in Britain and Central Europe (Oden 1968). As well as a researcher, Oden was also a scientific entrepreneur who pursued an international campaign to inform the wider scientific community about the acid rain problem. Rather than turning to a scholarly journal, Oden chose to first publish an account of his findings in a prestigious Stockholm newspaper, *Dagens Nyheter*; this was the first public airing of the acid rain problem (Gould 1985: 16).

At about the same time, a University of Toronto fisheries zoologist, Harold Harvey, and his graduate research assistant, Richard Beamish, were also finding evidence of fish deaths in lakes in Killarney Park near Sudbury. Like Oden, Harvey was an entrepreneur and took his research to three national magazines, but they all rejected it. Eventually, however, the story was published in July 1971 in the *Globe & Mail* which is distributed across Canada.

As Harrison and Hoberg (1991: 25) have observed, while science and technology can be important driving forces in spotlighting environmental problems, they are 'best considered necessary but not sufficient conditions for the emergence of issues'. Despite the rapidly developing critical mass of research on acid rain and its effects, it did not fully emerge as a public policy concern until the United Nations Conference on the Human Environment in Stockholm. This was the first major international forum to consider measures relating to the protection of the environment and the Swedes moved the case of acid rain prepared by their scientists front and centre, likening transborder pollution to a 'form of unpremeditated chemical warfare'. While no regulatory action was actually taken, the problem was publicly identified, named and politicised for the first time.

One tangible result of this conference was the initiation of a Norwegian research project entitled 'Acid precipitation – effects on forest and fish'. This SNSF Project, as it was called, was the largest multi-disciplinary study in Norwegian history and subsequently produced two major international scientific conferences in 1976 and 1980 (Regens and Rycroft 1988).

In June 1974, the acid rain debate spread from Scandinavia to the United States when Likens and Bormann finally published a report of their work in the widely read journal *Science*, which suggested that acid rain was falling on areas of the Eastern seaboard and that the sources of this precipitation were toxic emissions from coal-fired utilities and industrial plants. Within the year, the mainstream media picked up on the story. The *New York Times* (13 June 1975, p.1) reported that the increased acidity of rain was linked with reduced forest growth in New England and Sweden, with increasing lake acidity in Canada, Sweden and the United States and with corrosion damage to buildings. From this point on national media coverage accelerated, peaking in 1980 when there were twenty-two articles on acid rain indexed in the *Reader's Guide to Periodical Literature*, as well as numerous newspaper reports (Mazur 1985: 124–5).

Within US bureaucratic agencies, however, acid rain had difficulties emerging as an environmental problem with its own distinct identity. In fact, until 1980 acid rain was viewed as a subset of broader energy and environmental concerns, notably the controversy surrounding the implementation of the Clean Air Act of 1963 (Regens and Rycroft 1988: 118).

The acid rain problem, then, was initially assembled within the scientific community and brought to the public realm by Swedish officials in the Ministries of Agriculture and Foreign Affairs. While media interest was not by itself sufficient to move the incipient problem on to the policy agenda in the United States (and similarly in Britain), a continuing stream of articles in generalist scientific journals (*Science, Nature*), and magazines (*Environment, National Geographic*), national circulation news magazines (*Time, Newsweek*) and prominent daily newspapers (*New York Times, Globe & Mail*) did help to build public awareness and to distinguish the problem from the broader environmental problem of air pollution.

PRESENTING THE CLAIM

As a potential environmental claim, damage from acid rain faced several formidable obstacles. Unlike the pesticide problem which emerged just prior to acid rain's appearance on the world stage in 1972, the implied threat to human health was never as strong. Despite Smith's early attribution of deaths in Glasgow to acid sulphates and other air pollutants and the 'killer smogs' in Pennsylvania and London, in the late 1940s and early 1950s, it had proven very difficult to quantify the long-term health effects of low levels of acid haze (Gould 1985: 77). And unlike pesticides, you could not readily gauge residue levels in food, although acid inputs into lakes and streams and in the soil could be directly measured. Furthermore, the impact of acid rain was most deeply felt far beyond the populated urban areas where most people lived, concentrating in pine forests and pristine lakes. In this respect, it was quite different from the toxic chemical pollution which was situated literally in the backyards of city dwellers in areas such as the Love Canal neighbourhood. Finally, rather than a uniform phenomenon, acid rain was geographically spotty, seemingly devastating some areas and leaving others bare. Source countries and regions were at first little inclined to worry much about the problem as did receptor (victim) areas, especially since taking action entailed considerable economic sacrifice.

Given these obstacles, it was necessary to frame the acid rain problem in dramatic, even apocalyptic terms. The phrase 'acid rain' itself fits this requirement quite well. In popular American culture, acid conjures up images of motion picture thrillers in which deranged murderers are revealed to have turned against humanity after being facially scarred for life by corrosive acids which have been hurled at them. Of course, the harmful effects of acid precipitation were long term rather than immediate, but this distinction was often blurred in popular and media discourse. Gould (1985: 18) cites the examples of cartoons showing an umbrella shredded by acid rain and advertisements for a process which would allegedly 'acid rain-proof' the family car. Not surprisingly, Norman Quinn, a biologist who works in Canada's Algonquin Park, reports that visitors assume that the lakes there are acid pools and ask if it is safe to brush their teeth (1994).

In Canada and in the New England states, the aquatic effects of acid rain gave rise to the politically powerful 'dying lake' metaphor (Macdonald 1991: 245). In a similar fashion, the ravages of *Waldsterben* (the dying of the forest) were widely publicised in West Germany in the early 1980s. Especially influential was a widely read November 1981 cover story in the newsmagazine *Der Spiegel*, which depicted ailing conifers in front of smoking chimneys with the caption *'Der Wald Stirbt'* ('the forest is dying'). It was further predicted that large tracts of German forests would be dead within five years (Boehmer-Christiansen and Skea 1991: 189). These 'rhetorical motifs', as Ibarra and Kitsuse (1993) have termed them (see Chapter 2), did much to imbue what first appeared to be a largely technical-scientific problem with a distinct moral significance.

Acid rain claims-makers were also successful in linking it to some powerful cultural symbols. Despite the fact that the major source of damage to such priceless artefacts as the Pantheon and the Statue of Liberty appears to lie elsewhere, environmentalists frequently referred to these symbolic structures because of their emotional appeal (Regens and Rycroft 1988: 50).

Another effective rhetorical strategy was to link acid rain to the decline of the maple syrup industry in Eastern Ontario, Quebec and the New England states. Maple syrup summons up nostalgic feelings for the rapidly disappearing rural way of life and is a popular (and expensive) staple in farmers' markets and gift shops throughout this region. Although the claim that increases in the dieback of sugar maple trees were linked to acid rain damage originated in 1978 with maple syrup producers in the eastern townships of Quebec and later (1984) with Ontario producers from the Muskoka area, it was readily taken up in mass market publications such as *Canadian Geographic*. While subsequent research studies seem to indicate that this damage to the sugar maples was as much due to a host of other factors – attack by gaseous pollutants, notably ozone, root infections, drought, tent caterpillar infestation, general stress – as it was from the effects of acid rain (Borie 1987; McLaughlin 1985), the symbolic linkage between acid rain and the threat to maple syrup nevertheless remained part of the popular discourse on this environmental problem.

A further natural symbol which was tied into the acid rain problem was the common loon. The loon with its plaintive call is the favourite bird of many Canadians who go camping or to cottages on Northern lakes. Consider, for example, this letter to the popular 'discovery' magazine *Equinox* in which a Quebec reader describes his 7-year-old son's first encounter with a loon on a family fishing trip:

> This was a first call for my younger son; the look of total wonderment on his face is one that I will never forget. I removed a looney [a nickname for the Canadian dollar] from my pocket showing him the image on the coin and pointing to the loon on the water. I had to laugh at his expression when realization set in.
>
> (*Equinox*, Nov/Dec 1994, p.6)

Robert Alvo (1987), a Toronto biological consultant, found evidence from the 'Sixteen Ontario Lakes loon survey' and from provincial government documents to indicate that loons were raising very few chicks on acid lakes. His own two-year observational study of thirty lakes on which loons had nested in 1982 seemed to confirm this linkage. Alvo reported his results in 1987 in *Canadian Geographic* in an article entitled 'Is the laughter of loons to be stilled on acid lakes?' As Bosso (1987: 80) has noted in his analysis of pesticide poisoning as a public issue, the dramatic images of dead birds evoked by Rachel Carson's book *Silent Spring* was much more compelling to the media and public opinion than laboratory evidence which indicated that pesticides might cause harm somewhere down the road.

Acid rain claims-makers further provided 'grounds' for its importance and range by making a number of estimates of the magnitude of the problem (Best 1987). Among these estimates are the following:

- Rain falling on parts of the Eastern United States has increased in acidity to 100–1,000 times normal levels.

(Likens and Bormann 1974)

- Half of Quebec's 48,000 lakes are in danger due to acid rain.

(Chiras 1988: 360)

- If acid rain and snow continue to fall at current rates, 60 per cent of all the rivers in Sweden will reach a lethal level of acidity within 25 years and 90 per cent will attain that lethal threshold by the year 2050.

(Oden 1976: 21)

Collectively, these estimates of the damage induced by acid rain across a variety of ecological settings in a number of different parts of the world seemed to indicate that the problem was both extensive and was continuing to grow in magnitude. While less apocalyptic than some other subsequent environmental threats such as global warming and ozone depletion, acid rain was none the less frequently depicted in calamitous terms. One Canadian parliamentary report even went so far as to brand it 'the greatest threat to the North American environment in the recorded history of the continent' (Canada 1981:2), while the Royal Commission on Environmental Pollution in the UK labelled acid rain 'one of the most important pollution issues of the present time.

Finally, the acid rain problem was consistently warranted as a 'jurisdictional rights' issue. In Canada, this was supported by the widespread perception that Canadians were victims of pollution which originates from industrial sources in the United States. In the latter country, the transfer of air pollution from midwestern states to those in the East provoked similar tensions between source and receptor. In Northern Europe, the Scandinavian countries blamed Britain and the Ruhr valley in Germany for the increasing acidification of its surface waters. In each case, political action was justified on the basis that the polluter state or country was impinging on the environmental space of the victim and that this violated its sovereignty.

CONTESTING THE CLAIM

From the time it first appeared on political agendas in the early 1980s, the claim that acid rain was a serious environmental problem was vigorously contested in source nations. In the United States, the 'anti-control alliance' consisted of coal-burning electrical utilities, the coal-mining and auto industries, representatives of polluter states, the coal-mining unions, some state energy agencies, corporate lobby groups and industry-supported research institutions (Yanarella 1985: 42). In Britain, the major countering voice was the Central Electricity Generating Board (CEGB) whose twelve largest coal-fired electrical power generating plants were responsible for two-thirds of the country's sulphur dioxide emissions (Wilcher 1989: 37).

Environmental-based claims about acid rain have been especially open to contestation because they have been cultivated in 'a unique blend of scientific uncertainty and political misunderstanding' (Park 1987: 53).

Much of the scientific evidence concerning the environmental impacts of acid rain is circumstantial rather than conclusive. Even when addressing the acid rain damage to surface waters and fish populations, the one point where there is almost complete agreement among scientific assessments, it is difficult to quantify the extent of current damage, to predict future damage and to explain why some aquatic ecosystems are more vulnerable than others (Schmandt et al. 1988: 52–3). Rain often appears to function interactively, amplifying other environmental stressors but not always showing up as a direct cause itself; this can be seen in the case of sugar maple tree damage. Finally, the negative effects of acid rain are highly variable, appearing to depend on details in local geology, climate and vegetation (Schoon 1990, cited in Yearley 1992: 108). Opponents of acid rain regulation have consistently taken advantage of these strands of scientific ambiguity, using the need for a more detailed scientific assessment in order to delay the implementation of emission controls and other mitigative measures.

Politically, the acid rain issue is equally complex. Attempts by receptor countries to press source nations to adopt tighter emission controls have frequently been interpreted (or misinterpreted) as threats to the former's national sovereignty. Furthermore, the motivation of environmental claimants has been questioned by opponents on the grounds that they are acting more from selfish instrumental motives than from altruistic, moral ones. For example, some American opponents of acid rain have detected a 'Canadian conspiracy' whereby pressure on American utilities to adopt costly control strategies will allow Ontario Hydro to undersell their American counterparts in Eastern US markets (Schmandt et al. 1988: 133).

This political complexity is further enhanced in the United States due to the separation of powers between Congress and the President, the power of economic interest groups and divisiveness created by the existence of strong regional interests (Wilcher 1989: 61).

In considering the contestation of claims about acid rain, it makes the most sense to separate the discussion into four case studies: Scandinavia, Canada/United States, Germany and Britain. While these cases are by no means mutually exclusive, the patterns of claims-making and the policy stances which were taken in response to these claims varied considerably, with quite different outcomes.

Scandinavia

As we have already seen, the initial assembly of the acid rain problem occurred in Sweden in the period 1968 to 1972. Inspired by Oden's finding that the increased acidity of lakes in Southern Scandinavia and the attendant mortality of fish were linked to airborne sulphur emissions from Germany and Britain, Swedish scientists and diplomats were the first to focus world attention on the problem when they raised the issue at the 1972 United Nations Conference on the Environment. It is important to note that Oden's findings were not totally conceived in a scientific or political vacuum. A decade prior to Oden's discovery, the International Meteorological Institute in Stockholm had established the European Air Chemistry Network (EACN). EACN had compiled data which revealed that a core of highly acidic precipitation was falling over an area the size of the Benelux (Belgium–Netherlands–Luxembourg) countries.

With more than 80 per cent of sulphur emissions originating outside the country, Sweden (and Norway) had no choice but to press their claims in the international arena. They did this in three ways.

First, Scandinavia became the centre for acid rain research attracting scientists from across the world. Most prominent researchers who studied this problem (Eville Gorham, George Hendrey, Gene Likens, Richard Wright) went to Scandinavia to gain firsthand knowledge of acid rain and its damage (Ostmann 1982: 71). The SNSF project in Norway provided seven years (1972 to 1979) of detailed data on the effects of acid rain on forests and fish, establishing the basic parameters of knowledge for researchers elsewhere.

Second, Sweden, in its role as an environmental leader, followed up its 1972 initiative a decade later by hosting a Ministerial Conference on the Acidification of the Environment in June 1982. While most of the major source countries – Britain, United States, Poland, Czechoslovakia, the

Soviet Union – did not attend, the meeting was significant because West Germany, a large sulphur dioxide producer, actively supported the Scandinavian initiative for the first time and pledged a 60 per cent reduction from power stations and large factories within a decade.

Third, high profile campaigns against acid rain were organised in both Norway and Sweden. The Swedish campaign not only sought to influence domestic opinion but also targeted publics in other nations. Among other measures this included a leaflet on acid rain aimed at foreign tourists visiting Sweden, a four-colour 'Noah's Ark' poster (arguing that if acidification is allowed to continue we will soon have to build a new ark) printed in five languages and a multi-lingual Swedish film about acid rain, *Another Silent Spring*, which won prizes at large international film festivals in Czechoslovakia and the Netherlands (Park 1987: 173). Together with the Stop Acid Rain Campaign, Norway, the Swedish NGO Secretariat on Acid Rain publishes an English language tabloid, *Acid News*, with contributors from other nations who monitor government policies, scientific studies and other related developments.

Sweden and Norway are representative of countries where the populace is environmentally sensitive, acid rain is a high profile policy issue and there is widespread societal support for the government's commitment to reduce sulphur emissions to a level beneath 10 per cent of the total load (Wilcher 1989: 48). Acid rain claims have not been sharply contested here in large part because the economic interests of industry and government are more closely linked to those natural resources, i.e. forests, lakes which, it is believed, are the chief recipients of acid rain damage.

While concern with acid rain has continued in Scandinavia, there have been some recent changes in attribution. In their report, *Strategy for Sustainable Development*, the Swedish Environmental Protection Agency (SNV) concluded that the part played by forestry methods needs to be given greater prominence, especially in Northern Sweden where poor silviculture management methods are said to acidify the soil to roughly the same extent as air pollution (enviro 1993: 22). In the Southern portion of the country, however, transborder deposition is still seen as the major cause of acid rain damage to forests.

Canada/United States

While the political dynamics of acid rain contestation differ considerably between Canada and the United States, it makes sense to discuss the two cases together on account of the strong bilateral nature of the issue. According to data reported in 1979 by the Bilateral Research Consultation

Group, 50 per cent of acid rain deposits in Canada originated in the US, while 15 per cent of US deposits came from Canada. While the United States preferred to keep acid rain off the bilateral agenda, Canadian input into American policy-making consistently kept it there, mobilising political support for action on acid rain control and holding up its own acid rain control programme as an incentive for reciprocal action (Cataldo 1992: 397).

While acid rain had first been brought to the attention of the Canadian public in 1971 with the *Globe & Mail* report on Harvey and Beamish's research, it was not until five years later that it began to have a major impact. In 1976 to 1977, the issue received its first prominent airing in a series of articles written by reporter Ross Howard for the *Toronto Star*, the largest circulation daily newspaper in Canada.

Soon after, environmental movement organisations turned their attention to acid rain. In November 1979, environmental organisations in Ontario, with the assistance of their counterparts elsewhere in Canada and the United States, organised a major conference in Toronto entitled Action Seminar on Acid Precipitation. Rather than the exchange of technical and scientific papers, its purpose was to plan political action (Macdonald 1991: 247).

Two years later, the Canadian Coalition on Acid Rain (CCAR), representing more than fifty conservation, environmental, recreational, tourist and other interests, was established and opened offices in Washington as an officially registered lobby group. As Macdonald (1991: 248–9) has observed, CCAR had a number of advantages not usually enjoyed by other environmental advocates. Approximately one-half of its $200,000 to $300,000 (CD) budget was provided by the Ontario and Federal governments who saw an advantage in funding an organisation which would stress the need for American action on acid rain. Its Board of Directors included representatives of elite groups such as the Muskoka Lakes Association whose membership was made up primarily of affluent cottage owners from Toronto. CCAR moved early and effectively into the use of direct mail techniques for fundraising and to stimulate letter-writing campaigns. While careful to never officially support any specific amendments to the Clean Air Act in the US (Schmandt et al. 1988: 130), its two Washington staffers, Michael Perley and Adele Hurley, learned how to penetrate and manipulate the political networks in American federal politics into becoming effective claims-makers in a city full of competing lobbyists. Not only did CCAR build coalitions on Capitol Hill but they also linked up with American environmental interest groups such as the National Clean Air Coalition (NCAC).

In the United States, the 'pro-control alliance' consisted of professional environmental groups, environmental research institutes, pollution control and resort industries, some state energy agencies, environmental and leisure lobbying groups and Canadian supporters such as CCAR and the Cana Embassy (Yanarella 1985:42). Entrepreneurial university-affiliated scientists in the US were also centrally situated and played a major role in pressing for a national monitoring network and programme of research. However, they tended to isolate themselves from the politics of acid rain in order to preserve their autonomy and protect themselves from failures to reach consensus over emission control policy (Zehr 1994).

The environmental position was spearheaded by the NCAC, a coalition of environmentalists, conservationists, health care groups, unions, consumer advocates and others. NCAC's political activities included testifying before Congressional committees, informing legislators of the merits of relevant positions and grassroots political education in the form of legislative analyses and the promotion of media coverage (Schmandt *et al.* 1988: 118).

Political analysts who have studied the acid rain issue in America have urged caution in assessing the role of these environmental claims-makers in bringing about legislative action on the control of acid rain (Wilcher 1989:80). Schmandt *et al.* (1988: 259) attribute this to their relative failure to mobilise public opinion. This reflects in part the regional character of the acid rain threat in the United States, in part the perception that the extent of damage is not serious enough to overcome opposition to new taxes or fear of lost jobs.

From the time Likens and Bormann's research was first reported in the *New York Times* in 1975 to the eventual signing of the Clean Air Act amendments in 1990, which required that sulphur dioxide emissions in the US be reduced by approximately one-half by the year 2000, fifteen years had passed. This included a period of virtual inaction during the Reagan administration in which coal-mining and auto industry interests were predominant. When legislative progress was finally made in amending the Clean Air Act strong lobbying by environmental interest groups did play a significant role, but only in combination with a number of political factors, notably, the waning influence of the midwestern coal and utility bloc and the ascendancy of a group of Congressional leaders who supported acid rain control legislation for a combination of regional economic and environmental health reasons (Cataldo 1992: 406). In contesting the acid rain problem, environmental claims-makers in the United States were able to assemble, articulate and animate the problem but they were consistently stymied in engineering a solution. By contrast, in Canada, where public

139

opinion was less divided by regional interests and polluters were more cooperative, emission reductions were negotiated earlier than in the US, although the actual start date of the cuts was consistently delayed. What this suggests is that the social construction of environmental problems must be viewed in concert with interest group politics, especially when considering the latter stages of the claims-making process in which the implementation of solutions is the predominant activity.

Germany

By all accounts, there was virtually no support for the fight against acid rain in Germany until the early 1980s. Then, very rapidly, acid rain became an issue which rose to the top of the political agenda. There were two reasons for this.

First, in November 1981, *Der Spiegel* published its sensational story on forest death, presenting biochemist Bernhard Ulrich's hypothesis that large tracts of German forests would be dead within five years as an established fact (Boehmer-Christiansen and Skea 1991: 189). This provoked widespread alarm not only among environmentalists but also from those with direct interests in preserving the forests, notably forest farmers and powerful hunting associations in Southern Germany.

A second major trigger was the growing influence and rising membership of the Green Party who transformed acid rain into a potent political issue. While they represented at best only one-tenth of the electorate, *Die Grünen* 'could raise the political salience[1] of environmental destruction sufficiently to make pollution abatement a matter for party political competition and administrative activism' (Boehmer-Christiansen and Skea 1991: 76–7).

By 1983, the issue had been embraced by a new environmental action group called Robin Wood, formed with a hundred members including former Greenpeace activists. In its aim to attract media interest to acid rain, Robin Wood planned a campaign which included attention-grabbing tactics such as occupying power station flue stacks in Berlin and Cologne and planting 'dead trees' in the town centres in the Baltic cities of Kiel and Bremen. Such public actions continued for several years. For example, in February 1985, environmentalists staged a mock trial in Essen of RWE – a large power company that generated 40 per cent of the electricity in the former West Germany. Evidently, the company was incensed, the media were interested, the authorities were confused and the public was amused (Park 1987: 184).

Starting in 1982, the West German government responded by taking a series of steps to reduce sulphur dioxide emissions. At the Stockholm Conference on Acidification, West Germany pledged a 60 per cent reduction by 1993. Domestically, large combustion power plants decided to comply with tightened emission requirements by retrofitting their coal-fired power stations with flue gas desulphurisation (FGD), a cleansing technology developed in Japan and the United States in the 1970s. With the exception of certain innovative pilot projects, almost all of the cost of this retrofit programme has been borne by the largely privately owned utility companies which have in turn passed on these costs to electricity consumers (Boehmer-Christiansen and Skea 1991: 285).

Britain

Unlike the other three cases which we have been considering, acid rain never really became a viable environmental problem in Britain. There are several explanations for this.

Despite the romantic legend of Robin Hood and Sherwood Forest, the forests in Britain have neither the symbolic value which they have in Germany nor are they important industrial and leisure resources as is the case in Canada and Sweden. With virtually no wilderness land remaining in the South and little acquaintance with or love of forestry monoculture, other environmental issues such as the disappearance of birds, hedgerows and wetlands have created more concern and pressure group activity in the UK than has acid rain (Boehmer-Christiansen 1991: 71). Since no cherished environmental assets were perceived as being at stake, forest die-back was never perceived as a crisis worthy of major public concern.

Second, what Park (1987: 243) terms the 'case not proven argument' has been used repeatedly and effectively by the successive Conservative governments which have been in power in Britain since 1979.[2] The publicly owned Central Electricity Generating Board (CEGB) has been a leading proponent of this view, questioning the scientific validity of the causal linkages between sulphur dioxide emissions and environmental damage. CEGB not only espoused this view but it actively promoted it; for example, CEGB produced a film which maintains that scientific evidence does not justify the excessive costs of cleansing sulphur emissions (Wilcher 1989: 49).

Third, the acid rain issue was consistently framed within political circles as a threat to Britain's sovereignty. For years, the Scandinavian countries continued to be highly critical of Britain in its role as an acid rain exporter. Both the Norwegian and Swedish embassies in London have been the nucleus of information campaigns against acid rain, providing educational materials such as kits, pamphlets, etc., to British schools. This was part of a grassroots campaign to raise consciousness about the effects of acid rain in Britain. Norwegian dissatisfaction with British policy on acid rain was openly expressed in 1985 when the Norwegian Minister of the Environment arranged a press conference to voice serious objections to the CEGB film which argued that scientific evidence did not justify the excessive costs of cleansing sulphur emissions (Wilcher 1989: 49).

Further pressure has come from the European Community whose powerful policy-making component, The Commission, has been critical of Britain's failure to make a specific commitment to acid rain reduction, especially its decision to opt out of the 30 per cent Club – a group of nations committed to reducing sulphur dioxide emissions from factories and power stations. Thatcher Conservatives have generally resisted these external pressures for action in the same manner as they have resisted initiatives leading to greater European unity, although they have been forced to make some concessions.

This is not to say that no attempt was made domestically to construct acid rain as an environmental problem. While scientific opinion was divided on the importance of acid rain, there were several key reports which accorded it a high priority. A 1982 report written by the Nature Conservancy Council, a statutory body responsible for providing national advice on the impact of pollution on protected fauna and flora, detailed a long list of ecological changes and damage believed to be associated with acid rain. The following year, the Royal Commission on Environmental Pollution issued a report which cited acid rain as 'one of the most important pollution issues of the present time'. Seven months later, the report of the House of Commons Select Committee on the Environment concluded that British power stations were one of the major causes of acidity in thousands of European lakes and recommended that Britain join the 30 per cent Club immediately.

The Stop Acid Rain UK campaign was run by a coalition of scientists, environmental movement organisations (Friends of the Earth, Greenpeace, the Ecology Party, the Socialist Environment and Resources Association, Young Liberals Ecology Group) and sympathetic politicians (Park 1987: 7). Friends of the Earth, in conjunction with the UK World Wildlife Fund, was the first to issue a report, the *Tree Dieback Survey* (Rose and Neville 1985), which claimed that acid rain was actively causing damage to Britain's forests; opponents responded by branding the survey as scientifically flawed and Friends of the Earth as scaremongers. In addition to its sponsorship of scientific research, Friends of the Earth staged demonstrations, participated in the annual International Acid Rain Week and distributed public information materials (Wilcher 1989: 40). Despite considerable publicity generated by the British and the international Stop the Acid Rain Campaign(s), this environmental action had 'no perceptible effect on U.K. policy' (Boehmer-Christiansen and Skea 1991: 210).

Eventually, CEGB decided to retrofit some of its coal-fired power plants with FGD technology, although it was only willing to do so during scheduled routine maintenance times which take place every three years.

In June 1988, Britain agreed to cut its sulphur dioxide emissions from existing plants on an escalating scale over fifteen years, reaching a figure of 60 per cent below the 1980 level by the year 2003. This was less than was desired by their EEC partners and by environmental advocates but nevertheless still constituted a notable reversal in direction. While it is not possible to gauge precisely why this policy shift occurred, it appears that it reflected a combination of persistent attention from the House of Commons Select Committee on the Environment and aggressive external pressure from the EEC and Scandinavia which eventually came together to make a limited response politically necessary. Unlike the German case, the role played by environmentalists and by the mass media was at best indirect.

By 1993, Britain had achieved cuts of only 25 per cent in sulphur dioxide emissions, much lower than the cuts in the Netherlands and Belgium (50 per cent), France (60 per cent) and the former West Germany and Austria (70 per cent), Further cuts have been delayed by the privatisation in 1990 of the electricity industry and by the government's decision to slow down the licensing of 'clean' gas-fuelled power stations (Pearce 1993). Acid rain can thus be judged to have failed to make much progress as an environmental problem on the British political agenda.

CONCLUSION

As Solesbury (1976: 392) has pointed out, commanding attention for an environmental issue and demonstrating its legitimacy may put an issue on the political agenda but it will not guarantee that action will be taken. As these issues move towards the point of decision, they become increasingly confined to the decision-making institutions of the political system.

In the 1970s and 1980s scientific and environmentalist claims-makers were reasonably successful in distinguishing acid rain as an environmental problem distinct from air pollution in general and bringing it to the attention of the mass media. Once in the political arena, however, it succumbed to a variety of political constraints from the ideological rigours of Reaganism and Thatcherism to the pressures from utilities and industrial polluters. These claims-makers were especially handicapped by the uncertainties of acid rain research. By expanding the scientific case against acid rain from its effects on surface waters to allegations of forest damage, environmentalists inadvertently opened the door to the 'case not proven' argument since the latter claims were on shakier scientific ground than the former.

The one anomalous case is that of the Federal Republic of Germany. Whereas acid rain plodded along as a political–environmental issue in the UK and America for the better part of a decade, in Germany it went from the discovery stage to that of decisive government action in less than three years. Boehmer-Christiansen and Skea (1991) have argued that it was the conjunction of six contextual factors: a cultural and historical predisposition towards air pollution control; the widespread impact of deep green thinking and politics during the 1980s; a decentralised, proportional political structure allowing greater political representation to third parties; an economic management style which supported the stimulation of environmental investment through regulation; an energy sector structure which promoted nuclear power; and a proactive, legalistic, air pollution control system which accounted for this phenomenon. The German case is all the more remarkable because it arose from the forest death 'crisis' which was arguably the least viable of the scientific claims about acid rain damage.

In recent years, the acid rain problem has yielded to another generation of global environmental problems, notably those related to global climate change.[3] This is consistent with Hilgartner and Bosk's 'ecological' model in which incipient social problems compete within public arenas for access, legitimacy and action. In the Spector–Kitsuse natural history model, mature social problems inevitably spawn alternative or counter institutions when it becomes clear that bureaucratic organisations set up to deal with perceived problems are not aggressively and effectively dealing with the situation. Despite gaps in existing sulphur dioxide emission programmes and the relative failure internationally to control nitrogen oxide emissions, this seems unlikely to happen in this case.

If the problem is re-articulated in the future it will most probably be in conjunction with the air pollution problems of countries outside the 1979 Geneva Convention on Long Range Transboundary Air Pollution, notably China, South Africa, Brazil and Venezuela. Another possibility is that acid rain concern will be revived as part of a broader campaign against urban photochemical smogs. The emphasis here would be less on sulphur dioxide emissions from utilities and more on the junior partner in the acid rain problem – nitrous oxide from cars and other fossil fuel burners (Pearce

1990: 60). Finally, acidification could be revived in relation to its role in reducing the biodiversity of plant and animal species. This is thought to be particularly the case in microenvironments suffering from droughts caused by the greenhouse effect where a sudden rainstorm brings a kind of 'acid pulse' which is deadly for certain species (Pearce 1990: 60). In Sweden, it has been estimated that in about one-fifth of the country's lakes, species diversity has already been reduced by between 10 and 20 per cent (enviro 1993: 23). With biodiversity loss as one of the major global environmental problems in the international community today (see Chapter 8), this could prove to be a significant way of reformulating the risks posed by acid rain.

8

BIODIVERSITY LOSS

The successful 'career' of a global environmental problem

Along with global warming, the conservation of biodiversity was one of the two major issues at the June 1992 United Nations Conference on Environment and Development (UNCED) in Rio de Janeiro. It has been called the 'hottest' environmental topic of 1993 (Mannion 1993) with a burgeoning academic and popular literature devoted to exploring its parameters. Yet twenty years ago, the term biodiversity was unknown and it was not to be found in any compendium of threats to the environment. The skyrocketing career of biological diversity loss is a good illustration of how a 'transnational epistemic community' (see Chapter 4) can assemble, present and successfully contest a global environmental problem.

As a concept, biodiversity is multi-layered with various levels of meaning (Udall 1991: 82). Officially, it has been defined as 'the variability among living organisms from all sources, including *inter alia*, terrestrial, marine and other aquatic ecosystems and the ecological complexes of which they are part' (Tolba and El-Kholy 1992). More simply, it is an umbrella term for nature's variety – ecosystems, species and genes (Environmental Conservation 1993: 277).

Biodiversity is generally acknowledged to exist at three distinct levels: ecosystem diversity; species diversity; and genetic diversity.

Ecosystem diversity refers to the variety of habitats that host living organisms in a particular geographic region. This variety is said to be shrinking in the face of accelerating economic development. Udall (1991: 83) uses the metaphor of a ripe pumpkin which has been hollowed out to describe the damage to our ecosystems which has been inflicted by trapping, ploughing, logging, damming, poisoning and other forms of human intrusion. With the rapid pace of development, land ecosystems are described as increasingly taking the form of 'habitat islands'; for example, a patch of tropical forest surrounded by croplands (Franck and Brownstone 1992: 37).

146

Species diversity refers to the variety of species that are found in an ecosystem. While there have been notable episodes of species extinction in the past, the scale of loss today is judged to be unprecedented in the history of humankind (Lovejoy 1986: 16). Much of this is attributable to loss of ecosystem diversity; as a broad general rule, reducing the size of a habitat by 90 per cent will reduce the number of species that can be supported in the long run by 50 per cent (Tolba and El-Kholy 1992: 186).

Genetic diversity refers to the range of genetic information coded in the DNA of a single population species. Biologists value genetic diversity because it is seen as the basis for permitting organisms to adapt to environmental change. For example, in agriculture, wild strains of plants are valued because they often contain genes that are vital in fighting off pests or disease, unlike domesticated 'monocultures' which are much more vulnerable. In the animal world, inbreeding among a population stranded by habitat loss or commercial exploitation leads to an inability to survive in the long term; for example, this is the situation of the grizzly bears in Yellowstone Park in the American West (Udall 1991: 82).

When all three levels are viewed together, biodiversity loss appears to be a newly minted environmental problem. However, as Barton (1992: 773) has observed, there have long been a variety of treaties governing individual elements such as the international trade in endangered species, regional conservation and the conservation of particular species. For example, the Migratory Birds Convention, signed in 1917 by the United States and Canada, was a key piece of legislation in the campaign during the first part of this century to save North American birds. And, in 1911, six years earlier, a major international agreement, the Convention for the Protection and Preservation of Fur Seals, had been signed.

BACKGROUND FACTORS

There are three major developments which set the stage for the rise of biodiversity loss as a major environmental problem in the 1980s.

First, the growing economic importance of biotechnology meant that a greater financial value was increasingly being placed on genetic resources, a value that was recognised through intellectual property rights (Barton 1992: 773). Of special importance here was a landmark decision by the US Supreme Court (*Diamond v. Chakrabarty*) that allowed for the first time the patenting of a genetically engineered microbe, in this case an oil-eating bacterium developed by a General Electric research scientist named Ananda Chakrabarty. Also of significance was the passage a decade earlier of the US Plant Variety Protection Act (PVPA) which set up a patent-like system to govern the seed industry under the auspices of the US Department of Agriculture rather than under the more rigorous requirements of the US Patent Office. These events were significant for two interrelated reasons.

By raising the monetary stakes involved in the development of genetic resources, a conflict was fanned between the developed nations who wished to ensure open access to plant and animal genes and the less developed nations in which the bulk of these genetic materials were to be actually found. The latter began to see the genetic prospecting of the multinational pharmaceutical and chemical companies headquartered in Northern nations as a form of 'plundering' for which compensation should be paid.

At the same time, genetic diversity also became an international development issue due to the entry of several well-known rural activists (Cary Fowler, Pat Roy Mooney) to the debate over plant patenting. Fowler, a farmer from North Carolina, had worked with food activists Frances Moore Lappé and Joe Collins on the national bestselling book, *Food First*, an indictment of the world food system. Fowler became a one-person lobby opposing changes to the seed patent laws. In the 1979 debate over a proposal to amend the PVPA so as to add six 'soup vegetables' theretofore excluded from the act, Fowler

> turned his mailing list loose on Congress, went to the Press, wrote articles about the issue, and travelled around the country alerting other groups to the 'seed patenting' issue. Fowler rallied scientists and church interests and wrote to the secretary of agriculture, Bob Bergland, urging him to consider the impact of rising seed costs on small farmers.
>
> (Doyle 1985: 67–8)

Mooney, a Canadian from the province of Manitoba, helped to internationalise the seed issue both by his participation in a network of activist scholars working on Third World issues and through his widely circulated paperback book, *Seeds of the Earth*, published in 1979 for the Canadian Council for International Cooperation and the International Coalition for Development Action.

Second, the emergence of conservation biology in the late 1970s as an academic specialty provided a nesting spot for research on biodiversity. Conservation biology is an applied science which studies biodiversity and the dynamics of extinction. It differs from other natural resource fields such as wildlife management, fisheries and forestry by accenting ecology over economics (Grumbine 1992:29). The role of the conservation biologist is

148

to provide 'the intellectual and technological tools that will anticipate, prevent, minimize and/or repair ecological damage' (Soulé and Kohm 1989:1). Conservation biology is thus a 'crisis discipline'[1] which draws its content and method from a broad range of fields within and outside of the biological sciences.

Conservation biology was formally recognised as a discipline in 1985 with the creation of the Society for Conservation Biology (SCB). Within three years, the membership of the Society had swollen to nearly 2,000 members (Tangley 1988: 444). SCB is significant because it has provided a central forum for the communication of knowledge about conservation and biological diversity, especially through its journal, *Conservation Biology*.

Another critical node in the development of the discipline was the establishment of the Center for Conservation Biology (CCB) in the Department of Biological Sciences at Stanford University in California. The Center's main activities are basic and applied research, education and the application of conservation biology principles to genetic resources, species, populations, habitats and ecosystems. CCB consults not only within the United States but also internationally, especially in Latin America (Franck and Brownstone 1992: 66), providing yet another link between research on biological diversity and the international development scene.

By the late 1980s, conservation biology had begun to develop rapidly at institutions of higher learning. A pioneering textbook, *Conservation Biology: Science of Scarcity and Diversity*, had been adopted by classes at thirty-seven US colleges and universities as well as overseas (Tangley 1988: 444). In 1985, the first conservation biology course was taught at the University of California at Berkeley with an emphasis on the biological foundations for conservation (Millar and Ford 1988: 456). While research funding was still modest, partly because of a perception that conservation biology was a 'soft' discipline, advocates of biological diversity as an environmental problem nevertheless had an increasingly powerful academic medium for spreading their message and for building a constituency.

Third, a legal and organisational infrastructure was being assembled in the 1970s within the United Nations and other non-governmental organisations (NGOs)[2] dealing with various elements of the biodiversity problem.

In 1971, the *Convention on Wetlands of International Importance especially as Waterfowl Habitat* was agreed upon with the dual purpose of designating environmentally sensitive areas for migratory waterfowl and facilitating transborder cooperation among countries situated along their travel routes. This agreement was staffed by a secretariat provided by the International Union for Conservation of Nature (IUCN).

The *Convention Concerning the Protection of the World Cultural and Natural Heritage* (held in Paris in 1972), prepared under UNESCO (United Nations Economic, Social and Cultural Organization) supervision, established exceptional World Cultural Sites such as Serengenti National Park in Tanzania, the Queensland Rainforests in Australia and Great Smokies National Park in the United States, some of which rated quite highly in biological diversity. The agreement established a world heritage fund to assist nations who may have difficulty in paying for the protection of these unique sites.

In 1973, the *Convention on International Trade in Endangered Species of Wild Fauna and Flora (CITES)* was proclaimed in Washington with a secretariat staffed by the United Nations Environment Programme (UNEP) located in Lausanne, Switzerland. This convention established lists of endangered species for which international trade is to be controlled via permit systems. CITES was limited, however, insofar as it was directed at individual species rather than the habitats in which they resided.

Finally, the *Convention on Conservation of Migratory Species of Wild Animals* (held in Bonn in 1979), with a secretariat furnished by UNEP, provided a framework for international cooperation among states which hosted animals whose travels regularly take them across national boundaries. A central aim of this convention was to coordinate research, management and conservation resources such as habitat protection and hunting regulation affecting migratory species.

These international legal agreements were supplemented by a number of regional measures; for example, conventions on the conservation of nature, natural resources, wildlife and natural habitats pertaining to the South Pacific (1976), Africa (1968) and Europe (1976) and by the designation of Biosphere Reserves under a UNESCO programme. Taken as a whole, such measures were not only useful in their own right as a means of fostering, if not enforcing, useful cooperation among nations in conserving biological diversity, but they also put into place a global system upon which more far-reaching and stringent international legislation to conserve biological diversity could be modelled. Furthermore, they established epistemic networks of research, communication and coordination which were vital in moving biodiversity along to its status today as a major environmental problem.

ASSEMBLING THE CLAIM

In contrast to those global environmental problems which involve damage by pollutants to the atmosphere (or stratosphere) – global warming, ozone depletion, acid rain – the threatened loss of biological diversity has been less dependent on the dramatic discovery of an alteration in nature; for example, the ozone 'hole' over the Antarctic or 'forest die-back' in the Black Forest. Rather, it has developed in the context of a steady outpouring of studies which have cumulatively raised the alarm bells.

Taken as a whole, these studies have often lacked precision. Estimates of the projected number of extinctions which might be expected by the end of the century have varied not only widely but wildly (Brown 1986: 448), ranging from hundreds of thousands to over a million. In other cases, the estimates have been made in terms of rates, a device which both implies a greater precision than is possible given current knowledge and which leads to some questionable figures. Most notably, the 'one extinction per minute' rate used by some authors is equivalent to 525,600 extinctions per year, an unlikely or impossible total about ten times the number usually cited (Lovejoy 1986: 14).

Furthermore, the enormity of the problem has meant that reliable information is difficult, if not impossible, to assemble. So little is actually known about how species interact in ecosystems, about how they depend upon each other and about how they recover from episodes of disturbance that 'actions required now to avoid future disasters must be undertaken without sufficient knowledge to make considered choices' (Norton 1986: 11). In the face of this scientific uncertainty, those who have promoted biodiversity as an environmental problem have fallen back on the 'precautionary principle' (see Chapter 5), suggesting that the wisest course is simply to avoid actions that needlessly reduce biological diversity (Tolba and El-Kholy 1992: 197).

How, then, were conservation biologists and other claims-makers able to elevate biodiversity loss to the status of a notable environmental problem, given the relative lack of authoritative research data on the subject?

Wilson (1986: v) believes that the rising interest during the 1980s among scientists and portions of the public in matters related to biodiversity and international conservation can be ascribed to two more or less independent developments.

The first was the convergence of data from three different areas of research – forestation, species extinction and tropical biology – such that global biodiversity problems were brought into sharper focus and thought to warrant broader public exposure. This critical mass of data was not sufficient to build an airtight case for immediate worldwide action but it did raise the profile of biodiversity to a level sufficient to provoke a stream of academic conferences, political hearings and public fora.

The largest of these was the National Forum on BioDiversity held in

Washington, DC, on 21–24 September 1986 under the auspices of the National Academy of Sciences and the Smithsonian Institution. This forum featured sixty leading biologists, economists, agronomists, philosophers and international development experts. The lectures and panels were regularly attended by hundreds of people and the final evening's panel was teleconferenced to an estimated audience of 5,000 to 10,000 at over a hundred universities and colleges in the United States and Canada. It was at this conference, Wilson (1986: vi) notes, that the term 'biodiversity' was first introduced by organiser Dr Walter G. Rosen, a programme officer from the Commission on Life Sciences, National Research Council/National Academy of Sciences. It is worth noting that in spite of Wilson's protests that the term biodiversity was too 'catchy' and 'lacks dignity', Rosen and other Academy staff members persisted on the grounds that the term is simpler and more distinctive, and therefore the public would remember it more easily (Wilson 1994: 359).

The second development was the growing awareness of the close link between the conservation of biodiversity and economic development, especially in Third World nations. This elevated biodiversity loss from a scientific environmental problem to a wider status as a sociopolitical problem. In the industrial nations of the North, destruction of tropical rainforests and other Third World habitats was decried on the basis that it threatened a vast untapped reservoir of species which could potentially prove useful in providing new foods, medical treatments and other products. At the same time, in the countries of the South, biodiversity loss was feared for its impact on local farmers and others whose livelihoods depend on the maintenance of traditional ecosystems. In time, these two objectives were to come into direct conflict but initially they acted in concert so as to reframe biodiversity loss as a 'development' problem of considerable importance.

This integration of conservation and development found a significant funding source in the US Agency for International Development (USAID) which expanded into the area in the 1980s by mandate from Congress. In addition to sponsoring individual projects and conferences in lower income countries, USAID administers a sizeable peer-reviewed research grant programme. The centrepiece of the latter is the Conservation of Biological Diversity project. This has two main components: cooperative funding of National Science Foundation (NSF) grants for research that contributes to the conservation of biodiversity in developing countries, and core funding for the Biodiversity Support Program, a consortium formed by the World Wildlife Fund, The Nature Conservancy and the World Resources Institute. USAID projects have been carried out in Latin America, the

Caribbean, Sub-Saharan Africa and North Africa as well as Europe and Asia (Alpert 1993: 630). At present, the agency invests about $100 million a year in biodiversity programmes around the world (Angier 1994).

The assembly of biodiversity loss as an environmental problem benefited greatly from the participation of several well-known scientific entrepreneurs or champions who were extremely active in promoting it both within and beyond the parameters of science.

The Ehrlichs, Paul and Anne, had already achieved a measure of fame in the late 1960s and early 1970s for their campaign to make overpopulation the centrepiece of the environmental crisis. Subsequently, the two biologists turned their attention to the problem of biodiversity loss. In 1986, they founded the Center for Conservation Biology at Stanford University with Paul Ehrlich as president (see above). In 1981, the Ehrlichs published *Extinction*, one of several high-profile books which appeared on the topic of endangered species and biodiversity around this time. Here they infused the biodiversity problem with a moral dimension using the Noah Principle to claim that the foremost argument for the preservation of all non-human species is the religious belief 'that our fellow passengers on Spaceship Earth . . . *have a right to exist*'.

A second major champion of the conservation of biodiversity was the celebrated Harvard entomologist, Edward Wilson. Widely known as one of the founders of the field of 'sociobiology', Wilson is also a 1978 Pulitzer Prize-winning author whose recent bestselling book, *The Diversity of Life*, has been carried as a selection by book clubs across the US and Canada. In his autobiography, *Naturalist* (1994), Wilson reports that he was 'tipped into active engagement by the example of my friend, Peter Raven', who had been writing, lecturing and debating the issue of mass extinction since the late 1970s. Among his contributions, Wilson edited a key collection of articles arising out of the 1986 National Forum on BioDiversity under the title *Biodiversity* (1988); this became one of the bestselling books in the history of the National Academy Press (Wilson 1994: 358).

Other key figures in assembling the problem of biodiversity loss were Raven, director of the Missouri Botanical Garden, Norman Myers, well-known international conservationist who published the book *The Sinking Ark* in 1979, and Michael Soulé, a founder and populariser of the discipline of conservation biology.

Longtime friends who had similar interests, moved in the same circles and often did fieldwork in the same areas (Mazur and Lee 1993: 703), Ehrlich (Paul), Raven and Wilson were involved in many of the same endeavours to promote biodiversity loss as a global environmental problem. Ehrlich and Raven organised and chaired panels at the 1986 forum in

Washington. In 1989, Raven and Wilson gave expert testimony before the US Senate Subcommittee on Environmental Pollution. Wilson and Ehrlich were contributors to the special biodiversity issue of *Science* in August 1991. And all three scientists were founding members of the Club of Earth, an activist group formed to bring scientific attention more quickly to important but neglected environmental problems (Brown 1986). Mazur and Lee observe of this trio:

> Their research productivity, their eminence and their social and institutional contacts gave them strong voices within the scientific establishment and good access to Federal and private sources of funding, which supported both their scientific and policy efforts.
>
> (1993: 703)

Wilson (1994: 357–8) refers to a 'loose confederation of senior biologists that I jokingly call the "rainforest mafia" whose members besides Raven, Ehrlich and himself included Jared Diamond, Thomas Eisner, Daniel Janzen, Thomas Lovejoy and Norman Myers and who were instrumental in advancing claims about the importance of biodiversity loss.

PRESENTING THE CLAIM

In presenting biodiversity as an environmental claim and keeping it on the public agenda, proponents face three formidable problems (McNeely 1992: 25).

First, unlike some other environmental problems such as toxic dumps or oil spills at sea, there is no easily identifiable opponent against which public opinion can be galvanised. Instead, the root causes of biodiversity loss are said to be found in basic economic, demographic and political trends including the relentless human demand for commodities from the tropics, runaway population growth and the escalating debt burdens of Third World nations (McNeely *et al.* 1990b).

Second, the loss of biodiversity has no immediate impact on human lifestyles in the First World nations where the resources which could be applied to acting upon the problem are concentrated. Indeed, with the exception of a small number of 'charismatic megafauna' – whales, gorillas, whooping cranes, bald eagles[3] – most threatened organisms consist of creatures such as fungi, insects and bacteria which most people would not hesitate to step on (Mann and Plummer 1992: 49). This problem is even more exaggerated at the system level where, as Noss (1990) has sardonically observed, 'you can't hug a "biogeochemical" cycle'.

Third, the collective benefits of taking action are notably imprecise. At most, conservationists can speculate that somewhere in the vanishing rainforest lies the cure for cancer or Aids but there are no iron-clad guarantees. By contrast, the costs of implementation are more apparent and onerous especially on the domestic front in developed nations. As a result, public attention often begins to lag when the visible costs seem to outweigh the immediate benefits. Public support can be further eroded by high-profile controversies such as that which occurred in the United States in the late 1980s and early 1990s over the fate of the Northern spotted owl.[4]

Claims-makers have addressed these difficulties by embracing a 'rhetoric of loss' (Ibarra and Kitsuse 1993). As can be seen in the box on p. 156, public statements by conservation biologists and other policy entrepreneurs stress that we are 'at a crossroads in the history of human civilization' (McNeely *et al.* 1990b: 40). Failure to act decisively is equated with turning down the road to chaos or driving a business into liquidation. Many of these metaphors are borrowed from the rhetoric of another environmental problem – overpopulation. Once again, we are depicted as rapidly approaching the 'limits of growth', thereby running the risk of surpassing the 'carrying capacity' of the planet. Lester Brown, president of the Worldwatch Institute, a well-known environmental think-tank, uses the rhetorical motif of a 'race' to describe how the momentum inherent in population growth with its attendant problems for biodiversity is pushing us rapidly towards a catastrophic finish line (1986).

The rhetoric of 'catastrophe' was further enhanced by linking it to the fate of the dinosaurs.[5] In 1980, two eminent scientists from the University of California at Berkeley, Nobel prize-winning physicist Luis Alvarez and his geologist son Walter, proposed that the dinosaurs had perished as the result of climate changes brought about by an asteroid which had crashed on earth sixty-five million years ago. Few scientific theories have attracted more public interest as quickly as did this controversial claim, a fact not lost on biodiversity activists who often used the dinosaurs as a point of comparison (Mazur and Lee 1993: 703). Indeed, this is still the case today, as evidenced by a recent television advertising campaign by the Humane Society of Canada which proclaims: 'it is the greatest extinction rate since the end of the dinosaurs. '

A subsidiary idiom here is that of 'entitlement'. Thus both Raven, in his testimony before the 1981 Congressional committee, and the IUCN in the introduction to *World Conservation Strategy* reiterate a memorable slogan to the effect that 'we have not inherited the Earth from our parents, we have borrowed it from our children'.

Running parallel to this 'doomsday' rhetoric is a second type of claims language which stresses the positive economic benefits of preserving

**A sample of rhetorical statements on biodiversity loss by
prominent environmentalists/scientists**

'We have not inherited the Earth from our parents, we have
borrowed it from our children.'

> Peter Raven in Congressional testimony (see Kellert 1986) and
> IUCN *World Conservation Strategy*, Introduction (1980)

'We are in a race. Maybe we should call it a contest.'

> Lester Brown (1986)

'We're treating the world as a business in the process of
liquidation.'

> Peter Raven, cited in Gooderham (1994a)

'The future well being of human civilization and that of many
of the 10 million other species that share this planet hangs in
the balance.'

> McNeely *et al.* (1990b)

'In the last twenty-five years or so, the disparity between the
rate of loss and the rate of replacement [of species and
populations] has become alarming; in the next twenty-five
years, unless something is done, it promises to become
catastrophic for humanity.'

> Ehrlich and Ehrlich (1981)

diverse habitats. Using warrants which are loaded with financial figures, proponents favour a 'rhetoric of rationality' (Best 1987). For example, Walter Reid, a vice-president of the World Resources Institute, writes about the 'economic realities of biodiversity':

> Currently some 25 per cent of U.S. prescriptions are filled with drugs whose active ingredients are extracted or derived from plants. Sales of these plant-based drugs amounted to $4.5 billion in 1980 and an estimated $15.5 billion in 1990. ... In Europe, Japan, Australia, Canada and the United States, the market value for prescriptions

and over-the-counter drugs based on plants was estimated to be $43 billion in 1985.

(1993–4: 49)

Significantly, this rhetoric uses the language of frontier development, for example, referring to 'bioprospecting' (Eisner 1989–90; Reid *et al.* 1993) or 'biotic exploration' (Eisner and Beiring 1994). It is suggested that somewhere in the 'biotic wilderness' scientists will find an equivalent of Madagascar's rosy periwinkle with its famous cancer-fighting properties (Eldredge 1992–3: 92).

This depiction of tropical rainforests as the cradle of tomorrow's pharmaceutical medicine has recently spread into the arena of popular culture. In the American motion picture *Medicine Man*, Sean Connery played a maverick scientist who discovered a cure for cancer among the canopies of the South American rainforest, only to have his research site flattened by the bulldozers of a road-building crew. And in *Day of Reckoning*, a 1994 action movie with a 'new age' flavour, an adventurer hunts for a rare plant with medicinal powers in the rainforests of Burma. As W. H. Hudson's romantic novel *Green Mansions* illustrated nearly ninety years ago, the human threat to the diversity of life in tropical ecosystems can make a compelling drama, with strong moral and spiritual overtones.

CONTESTING THE CLAIM

While individual countries undertook unilaterally to protect endangered species and habitats, it became obvious at least a quarter of a century ago that concerted global action on biodiversity loss required some type of coordinated multilateral agreement. In fact, an International Convention on Biological Diversity was first proposed in 1974 and active planning for such an accord began in earnest in 1983 (Tolba and El-Kholy 1992). This process culminated in 1992 with the preparation of a Global Diversity Strategy under the auspices of three agencies: the United Nations Environment Programme (UNEP), the World Conservation Union (the successor to IUCN) and the World Wildlife Federation.[6] In order to carry out the recommendations of this strategy, it was proposed that a Convention on Biological Diversity be put forward at the United Nations Conference on Environment and Development (UNCED) in Rio in June 1992. By all accounts, this convention was conceived in a medium of considerable controversy, especially with regard to the question of access to genetic resources in Southern hemisphere nations.

At the core of the treaty could be found a basic tension between two conflicting commitments. On the one hand, the proposers wished to provide a mechanism whereby the international conservationist community could directly intervene in situations where sensitive environmental areas with diverse biological resources were threatened, notably in the tropical rainforests of Brazil. On the other hand, target nations were not eager to lose their national autonomy and give up the right to make their own decisions, particularly with regard to development projects. As compensation for allowing outsiders to infringe their traditional national sovereignty, less developed nations wanted something in return, specifically, financial resources and the transfer of technology from the industrial nations of the North.

Furthermore, the Southern nations wanted to use the occasion to tighten up access to their genetic resources which theretofore had been more or less free to all comers. According to Article 15 of the Convention. nations were to have sovereign rights over their genetic resources and grant access only on mutually agreed terms and with 'prior informed consent'. Other provisions attempted to deal with some of the more potentially exploitative aspects relating to the appropriation of Third World genetic resources by multinational biotechnology companies. Increasingly, these firms have begun to prospect tropical habitats for unusual species of plants and animals, to 'borrow' key genetic material, bioengineer and patent it and then license the improved product back to the country of origin at a hefty profit. Accordingly, the South argued for access to the results and benefits of biotechnologies developed in connection with those genetic materials which have been exported, specifically in the form of continuing royalties and technology sharing.

Even at the pre-summit stage, a number of these points were contested. For example, an earlier draft of the convention had called for two global lists – a Global List for Biological Diversity and a Global List of Species Threatened with Extinction on [a] Global Level – which would have spelled out in priority fashion the commitments that were required of signatories to the treaty. However, during the final negotiations at Nairobi leading up to the Rio Summit, these references to global lists were removed, a measure that was strongly contested by many delegates including the leader of the French delegation who refused to sign the final act. Similarly, a provision which would have furnished free 'scientific access' to genetic resources in biologically diverse nations was dropped from the final convention (Barton 1992).

At the summit itself, the United States incurred the wrath of other participants by refusing to sign the Biodiversity Convention, even though 153 other countries did so and the Secretary of the Environment himself was in favour. This appeared to be the result of considerable pressure on President Bush from American biotechnology trade associations, which objected to the provisions that would have meant that US firms must pay continuing royalties and share new patents and technological secrets with nations whose biological resources are the source of new products (Susskind 1994: 182).

The Biodiversity Convention was contested by a third party who was not present at the negotiations or the Summit. This was a coalition of farmers, ecological activists and others from Third World nations who felt that local people had been excluded from the formulation of the treaty, especially the provisions relating to intellectual property rights. The best-known spokesperson for this movement is the Indian ecofeminist, Vandana Shiva, an associate editor of *The Ecologist*.

Shiva and her movement have attempted to wrest 'ownership' of the problem of biodiversity loss from conservation biologists, non-governmental global environmental organisations and government negotiators who they accuse of assuming a mantle of leadership which is not theirs to wear. In particular, they object to the exclusion of the original donors of genetic resources – Third World farmers – from the exchange of resources and knowledge which the Convention governs. The basic problem, Shiva states, is that

> those 'selling' prospecting rights never had the rights to biodiversity
> in the first place and those whose rights are being sold and alienated
> through the transaction have not been consulted or given a chance
> to participate.
>
> (1993: 559)

Shiva observes that even in the case of the 1991 agreement between Merck Pharmaceuticals and INBIO, the National Biodiversity Institute of Costa Rica, a much-heralded and publicised example of how it is possible for multinational corporations to compensate the Third World for its genetic resources, the people living in or near the national parks in Costa Rica were not consulted, nor were they guaranteed any economic benefits. Rather, the agreement was forged between Merck and a conservation group formed at the initiative of a leading American conservation biologist Dan Janzen, who, it will be recalled, was a member of Wilson's 'rainforest mafia' (Shiva 1993: 559).

Opponents of 'commercialised conservation' (Shiva 1990: 44) have proposed the formulation of an alternative form of intellectual property, the *Samuhik Gyan Sanad* or Collective Intellectual Property Rights (CIPRs). These collective patents invest the right to benefit commercially from traditional knowledge in the community that developed it. Further-more, it is demanded that multinational companies seeking to utilise Third World genetic resources be compelled to deal through the village organi-sations who would hold title to these CIPRs. Failure to do so, it is claimed, would constitute 'intellectual piracy' (Shiva and Holla-Bhar 1993: 227).

CONCLUSION

Of the three environmental problems presented here, biodiversity loss has advanced the furthest in the international arena. In some ways this is surprising. While extensive, research on biodiversity largely navigates uncharted waters. Of the 1.4 million species which are presently known, only 5 per cent can be considered 'well known' and the relationships between many of them are a mystery (Gooderham 1994a: A-12). The theory which underlies ecosystem diversity is based primarily on small-scale studies of ponds projected on the larger screen of nature. The benefits of acting boldly are not precisely documented while the costs are consider-able. The ownership of the problem is disputed with multiple claimants.

Yet despite these drawbacks biodiversity promises to be 'the environ-mental theme of 1994' (Isaacs 1994: 17). There are several factors which account for this.

First, it is not purely an environmental problem but is simultaneously an economic and political question. For business, biodiversity has the potential to be made into a valuable resource which can generate a tidy profit. For Third World governments, it is both a source of foreign exchange and a window through which First World biotechnology can be accessed. For small farmers in India and other poor nations, it is a means of empowerment and resistance to the creeping power of global capital (Shiva and Holla-Bhar 1993).

Second, biodiversity loss constitutes a socially constructed environmen-tal problem which has brought together two well-established organisational sectors: the international development establishment and the global conservation network. Nested in a web of NGOs, it has an institutional momentum extending beyond that which is able to be generated by single environmental movement organisations such as Green-peace and Friends of the Earth which have more of an 'outsider' status.

Third, the biodiversity problem has not been constructed from scratch but has flowed out of the already long-standing problem of endangered species. The two problems are to a large extent symbiotic and synergistic. Biological diversity gives species endangerment and extinction a theoretical grounding which it previously lacked. The example of endangered species provides biological diversity with a specific focus and an emotional resonance which the more general issue often lacks. Furthermore, the preservation of diversity furnishes a rationale for action in rancorous environmental disputes such as those which have been raging over Great Whale River and the Clayoquot Forest in the Canadian North and West (Suzuki 1994b).

Finally, the location of biological diversity at the centre of the discipline of conservation biology means that, unlike acid rain, global warming and other more cross-disciplinary scientific problems, it has been buffered against the 'issue attention cycle' (Downs 1972) which affects a great many other environmental issues. Conservation biology provides biodiversity loss with a centre of gravity around which it can revolve, rotating out into the realm of international diplomacy and conflict but stabilised by the continual pull of research within this specialty area.

9

BIOTECHNOLOGY AS AN ENVIRONMENTAL PROBLEM

The bovine growth hormone conflict

In contrast to the two case studies which have been presented thus far, biotechnology as an environmental problem is situated in opposition to scientific entrepreneurship rather than arising out of scientific discovery. As a result, the opposition to biotechnology has been based to a greater extent on ethical and economic objections than it has on the marshalling of scientific evidence.[1] Furthermore, the risks associated with unregulated biotechnology are more 'hypothetical' than the here and now risks associated with some other environmental problems.

GROWTH OF BIOTECHNOLOGY

Biotechnology is an umbrella term for a wide range of techniques which are used to modify life forms for various research and commercial uses. Not only does it include 'genetic engineering' (recombinant DNA technology) with which it is often mistakenly equated, but it also covers many other common techniques, notably tissue or cell culture cloning, fermentation, cell fusion and embryo transfer.

It is often pointed out, especially by its proponents, that biotechnology is by no means something new, dating back to the use of yeasts in baking and brewing by the ancient Egyptians and Mesopotamians, and, in this century, to well-accepted methods of plant and animal breeding such as the production of hybrid varieties of corn. What is different about the new biotechnologies is the large number of genetic characteristics which can be manipulated, the speed with which this can be done, and the capacity (using recombinant DNA technology) to transfer animal genes to plants and vice versa.

While it has its roots in the Nobel prize-winning discovery of the double helix structure of DNA by Watson and Crick in the 1950s, contemporary biotechnology really dates its recent origins to the year 1973 when two California scientists, Stanley Cohen and Herbert Boyer, were able to isolate fragments of DNA in one bacterium and insert it into another. Commercially, the investment floodgates were opened after 1980 when the US Supreme Court's decision in the *Diamond v. Chakrabarty* case (see Chapter 8) gave a legal assurance that new bioengineered organisms would be able to be patented (Plein 1991: 476).

From 1979 to 1983, more than 250 small biotech firms were founded in the United States alone, while large, established chemical and pharmaceutical corporations such as Montsanto and Du Pont began significant in-house biotechnology research and development programmes from 1981 onwards. Lacking basic researchers with training in molecular genetics, major Japanese corporations gained access to frontier developments in the field by forming a variety of collaborative agreements with US biotech companies. In Europe, most new biotechnology firms were established in Britain, while on the Continent, large corporations and national governments were the driving forces, either linking up with American firms or creating US subsidiaries focusing on biotechnology (Fowler *et al.* 1988).

Krimsky (1991) has classified the environmental applications of biotechnology during its first decade (1980 to 1990) into five categories: (1) remediation of ecologically harmful technologies (pollution clean-up by micro-organisms capable of degrading oil spills, detoxifying chemical pollutants in soil and water and improving biodegradation in waste-water treatment plants, biological alternatives to agricultural fertilisers and pesticides); (2) improvement of existing plants (adaptability to harsh environmental conditions, herbicide resistance, growth acceleration); (3) modification of animals; (4) micro-organisms to mine the earth (oil recovery, microbial gold, copper and uranium mining); and (5) use of micro-organisms as 'biosensors' – markers in field tests and other environmental settings.

OPPOSITION TO BIOTECHNOLOGY

In America, the use of bioengineered micro-organisms to redesign plants so as to be more adaptable to harsh environmental conditions has raised the loudest alarm bells among environmentalists, largely as a result of the much publicised case surrounding the release of the 'ice minus' bacteria[2] into the environment. In Europe, the development of herbicide resistant plants has stirred up the sharpest controversy, especially in Germany.[3] At the halfway point of the second decade of biotechnological research, it has been the modification of animals, the topic of the case study featured in this chapter (pages 165–77), which promises to provoke the greatest degree of opposition among opponents of biotechnology.

Environmentalists in general have been somewhat hesitant to commit themselves to campaigns opposing biotechnological research and its applications in agriculture and industry. In large part, this lack of involvement reflects the fact that the negative effects of these biotechnologies are largely speculative and therefore do not provide any visible focus for protest in the present. In Britain, environmental groups have been somewhat more involved, although it cannot be said to top anyone's agenda.

In the United States, the issue of biotechnology and research safety 'piggybacked' the larger environmental movement in the 1970s with some of the major groups (Friends of the Earth, Natural Resources Defense Council, Environmental Defense Fund, Sierra Club) making representations at Congressional hearings and in the public arena in favour of more stringent research guidelines to ensure health and environmental safety (McAuliffe and McAuliffe 1981; Plein 1990). However, by the next decade, environmental opponents had moved to the sidelines of debate both due to the loss of resources during the Reagan presidency and to a rising notion among some environmental activists that biotechnology would make agriculture more efficient, thus reducing such damaging practices as intensive fertiliser and pesticide use and large-scale irrigation (Browne and Hamm 1990: 46).

Finally, institutional environmentalists, notably those associated with UNEP, have been exploring the use of biotechnologies in Third World countries to stimulate increased 'bioproductivity' in the form of elevated crop production and better pest control. UNEP has also been interested in the biodegradation of persistent pollutants and the biomining of metals (Fowler et al. 1988: 327–9). While due attention seems to have been paid to the importance of developing safety guidelines and to providing free access to information on gene technology, nevertheless, this practical interest in the economic application of microbial technologies has no doubt retarded any strong desire on the part of UNEP and its network of NGOs to place biotechnology on its shortlist of environmental issues which must be urgently addressed. Hindmarsh (1991: 203) has properly described the international Green Movement as devoting too few resources to mount an effective challenge – existing campaigns lack coordination and grassroots support is 'grossly inadequate'.

Consequently, opposition to biotechnology has come largely from a coalition of groups not normally central to environmental claims-making:

farmers, consumer organisations, Third World development animators, church task forces, animal rights groups, and so on. Plein (1990) classifies this condition as including two categories of groups: *conditional opponents* (agricultural groups, concerned scientists, environmental groups, public interest groups) who participate intermittently on an issue-by-issue basis, and *absolute opponents* who contest every issue. These two groups differ considerably in their use of strategy, with the former preferring political compromise and negotiation while the latter utilises the more confrontational tactics of litigation, publicity campaigns and public demonstrations.

RECOMBINANT BOVINE SOMATOTROPIN (bST): A 'BIO-BATTLE' FOR THE 1990s

Bovine somatotropin (BST), also known as bovine growth hormone, is a naturally occurring protein which is responsible for regulating the volume of milk production in dairy cows. Using a combination of recombinant DNA technology and fermentation techniques, researchers have developed a synthetic BST which can be mass-produced. Supplementation with this genetically engineered BST is claimed to benefit dairy farmers by boosting a cow's milk output by an average of about nine pounds (four kilos) per day, an increase of 10–25 per cent over herds where it is not used.

Recombinant BST (bST) has been approved for sale in the United States from the beginning of February 1994. A moratorium on its commercial use still remains in place in the countries of the European Community until 31 December 1999 and it has been banned in three Scandinavian countries. On the advice of its veterinary products committee, Britain refused licences in 1990 to Monsanto and Eli Lilly, American chemical companies with a bioengineered bovine growth hormone, due to doubts about its effects on animal health. In Canada, the product has been declared safe for human consumption by health and welfare officials but tests for animal safety and efficacy are still being conducted and analysed, with the dairy industry advocating a 180-day moratorium on licensing together with further research.

Despite their failure to prevent legalisation of bST in the United States, opponents continue to campaign against it, proposing a variety of measures from consumer boycotts to a written certification by milk suppliers that their milk is bST-free. This opposition was accented on 7 March 1994, a month and four days after approval, by a major advertising campaign in Canadian and American newspapers sponsored by The Pure Food Campaign under the headline, 'If You're Against Artificial Hormones And Antibiotics In Your Milk – You Better Act Now'.

Assembling the claim

At the centre of the North American campaign against bST has been the Washington-based environmental lobby group, The Foundation on Economic Trends (FET) headed by Jeremy Rifkin. Rifkin is often depicted as the leading foe of biotechnology. His 1983 book *Algeny* was the first popular critique of the science of genetic engineering, earning plaudits from a wide variety of sources from *Publishers Weekly* to the then Congressman Al Gore. Rifkin has been described as the 'czar of genetics litigation' (Krimsky 1991: 121). Working through FET, Rifkin launched a dozen lawsuits between 1984 and 1987 on issues which included transgenic (genetically manipulated) animals, the deliberate release of the ice minus bacteria, human genetic engineering and the construction of a biological defence laboratory in Utah. Unlike other high-profile environmental entrepreneurs such as the Ehrlichs, Barry Commoner and Edward Wilson, Rifkin's appeal does not flow from his scientific credentials and activities but rather from a combination of his skill in constructing legal challenges and his flair for attracting media attention. While some policy analysts have questioned the primacy of Rifkin's leadership, suggesting that he is effective at playing to the press but ineffective as a Congressional lobbyist (Plein 1991; Taylor 1988), there is no doubt that he has been the single most consistent and prominent opposing champion to biotechnology in agriculture and medicine.

In April 1986, a coalition of groups opposing the licensing of synthetic BST which, in addition to Rifkin and FET, included the Wisconsin Family Farm Defense Fund, the Humane Society of the US and Douglas La-Follette, the Secretary of State in Wisconsin, initiated what is by now an eight-year-old campaign by petitioning the US Food and Drug Administration (FDA) to prepare an environmental impact statement on the commercial production of bST.

Scientists in this case played virtually no part in assembling the claim. Rather, from 1968 to 1980, members of the scientific community constituted the 'bulwark of the pro-biotechnology community' (Plein 1991: 476). After 1980, when biotechnology came to be dominated by multinational corporations, academic researchers were drawn into a 'university–industrial complex' (Kenney 1986) in which the boundaries between public and private laboratories were often fuzzy. In some cases, university researchers made considerable fortunes working for the new biotech firms while still holding down university positions. For example, Herbert Boyer topped the *Genetic Engineering News'* 1987 list of 'molecular millionaires' with an estimated personal fortune of $88 million (Fowler *et al.* 1988: 181–2). Unlike conservation biology, molecular biology does not have as its mission the preservation of nature but rather its commercial manipulation. Furthermore, biotechnology proponents who are largely molecular geneticists or microbiologists specialise in biology at the molecular and cellular levels, unlike ecologists and conservation biologists who view biotechnology from a more holistic, 'organism–ecosystem–biosphere' level (Hindmarsh 1991: 201).

The case against bST was thus not centrally assembled and legitimated by science. On the contrary, agricultural and medical researchers almost uniformly asserted that genetically engineered bovine growth hormone was 'safe' for the cow and that the milk so produced posed no threat to consumers. For example, a panel of doctors and scientists assembled by the National Institutes of Health in the US reported unanimously that milk from cows injected experimentally with bST is safe for human consumption (*New York Times*, 8 December 1990: 1).

Presenting the claim

Plein (1990) has argued that public concern with the development of biotechnology has evolved in three separate stages from an emphasis on human ethics in the late 1960s and early 1970s to an interest in health and environmental safety concerns in the middle and late 1970s to a focus on economic issues in the 1980s. There is some evidence that this progression occurred in the controversy over bST. Opponents first framed the problem in terms of an ethical concern with a threatening new technology and later injected the economic issue of the declining family farm (Browne 1987). More recently, a concern with environmental health and safety has joined the economic argument at the centre of the anti-bST discourse.

The ethical opposition to the use of bST in the dairy industry is based on the assertion that biotechnology propagates a 'wrong view of nature' in which, as the German environmental organisation *Bund für Umwelt und Naturschutz Deutschland* (BUND) phrases it, 'the animal is degraded to a level of a machine' (Bora and Dobert 1992: 5). This means that humans have no right to manipulate and exploit other species as if they were robots. Rifkin (1983: 252) casts this in the form of a choice between two alternative futures. If we choose the path offered by bioengineering technology, we gain physical security but take a giant step down the path towards a sterile world full of biological replicas or facsimiles. By contrast, if we choose the ecological option, we sacrifice this security in order to rejoin the 'community of life'; that is, we become the equals rather than the masters of the rest of the living kingdom. This is a view clearly compatible with the eco-philosophy of deep ecology and its emphasis on 'biocentric egalitarianism' – the principle that all things on the Earth have an equal right to live and blossom and reach their own individual forms of self-realisation. Synthetic BST technology is thus depicted as part of a 'Faustian bargain' in which the world will ultimately be ecologically diminished.

A second, somewhat narrower ethical concern promoted by the animal rightists claims that the use of bST will make dairy cattle subject to more production stress and therefore disease. While moral in tone, this argument does have some basis in science. Animal scientists have identified a number of stress-producing conditions which result in loss of immune function and subsequently in sickness in livestock. 'Shipping fever', for example, is endemic among beef calves which are transported long distances from feed lot to farm under stressful conditions. Rather than emphasise the economic disadvantages of this 'burn-out' syndrome,[4] claims-makers from the Humane Society and other animal rights groups highlight the 'suffering' which they believe hormone-stimulated cattle undergo (Fox 1986: 1247).

The economic objection to recombinant bST states that its commercial use puts small dairy farmers in jeopardy. A major reason for this is that increased milk production will drive prices down with the result that, in the United States, an estimated 25–30 per cent of dairy farms may disappear (Kalter 1985: 128). Such a decline has been of particular worry to dairy farmers in the state of Wisconsin. Once the dairy capital of America, in 1992 Wisconsin was surpassed for the first time by California as the largest dairy state. It is not surprising then that protest action against this product began in Wisconsin in 1986, subsequently spreading across the Midwest, New York, Pennsylvania and New England (Schneider 1990a: A-21). The implied linkage between the introduction of bST technology and the death of the family farm has thus been central to the rhetoric used by farm groups such as the Wisconsin Family Farm Defense Fund.

Less visible has been another economic argument against bST: that it will threaten the key marketing and governing institutions of the dairy industry (check-offs, marketing trade councils, cooperatives) by upsetting the structure by which production increases are introduced (Browne and Hamm 1990: 42).

The distrust of new agricultural innovations has a long history in the United States. For example, in the 1940s, Southern sharecroppers accurately foresaw that the mechanical cotton picker would mean the end of jobs for labourers in the cotton fields of the Deep South as production shifted to the West (Daniel 1986). In the case of bST, this economic threat is acknowledged even by some of those who dismiss other concerns. For example, a 1991 report by the Congressional Office of Technology Assessment concluded that while the hormone does not pose a fundamental danger to consumer and animal health, its commercial use will put greater economic stress on small dairy farmers. Thus Julie Bleyhl, a lobbyist with the 300,000-member National Farmers Union, strikes a central chord when she states that 'our main concern about the hormone is the economic implications ... we don't see any good coming from this' (Schneider 1989a: 1).

The environmental health and safety discourse centres primarily around the claim that cows injected with bST are more likely to be given larger doses of antibiotics which subsequently find their way into the food and waste chains. The main reason for this increased use of antibiotics is a greater than normal incidence of mastitis, an udder infection which may increase as the cow gives more milk. This claim has received a measure of scientific support from several sources. In its August 1992 report to Congress, the General Accounting Office noted that cows treated with the hormone had a higher incidence of mastitis and expressed concern that antibiotics used to treat this condition might be passed on to consumers through milk consumption. Similarly, James Armour, chairman of the British veterinary products committee, observed that they had found an increased incidence of mastitis in cows on some of the farms where field trials were conducted (Hornsby 1990: 4).

In addition, it is claimed that the ill-effects of bST may not be readily apparent right now but may show up at some point in the future. Kuchler *et al.* (1990) draw a hypothetical comparison with the case of DES, a synthetic oestrogen used as a growth promotant in beef cattle beginning in the 1950s which was made illegal by the FDA in 1979, largely on the basis that it had been linked to a rare form of vaginal cancer in women whose mothers had used it to prevent miscarriages.

The environmental health and safety discourse emphasises the 'purity' of food and the 'contamination' posed by new agricultural products and techniques such as the use of bST. As one industry representative observed, 'It's becoming a very emotional issue. ... Milk is pure and wholesome; it says motherhood and children and nurturing' (Gooderham 1994b: A-5). In its 1994 'Consumer Warning' advertisement, The Pure Food Campaign (its name is in itself indicative of the importance put on this image of healthy, 'natural' dairy products) features a photograph of a child eating breakfast together with the caption:

Was there a dose of artificial growth hormone (rBGH) in her milk this morning? Or antibiotics?

Contesting the claim

With a worldwide market estimated to reach as high as $1 billion per year, oppositional claims about the dangers of bST have been aggressively contested by the animal health industry.

Kleinman and Kloppenburg (1991) have depicted this as a discursive struggle in which biotech corporations increasingly intrude into the public sphere in order to create a favourable image for their new products and processes. Focusing specifically on the case of Monsanto Corporation, the world's largest investor in agricultural biotechnology, they demonstrate how advertising and promotion are used in order to shape the debate over the introduction of new biotechnologies.

There are three main discursive elements in the Monsanto promotional campaign.

First, technology is depicted as having its own trajectory not subject to human intervention. Historically inevitable, technology is said to be always beneficial and any opposition is both misguided and irrational. Biotechnologies such as bST are seen as ushering in a new golden era in which the fight against world hunger which was begun in the Green Revolution[5] of the 1960s will finally be won.

A second discursive element is the ideology of scientific expertise. This suggests that only scientific experts are capable of making decisions concerning new technologies and that public intervention is unwise. It is insisted that such decisions must be made on scientific merit alone and not on the basis of social or political criteria. There is no acknowledgement here that scientists have interests beyond objective truth-seeking. By contrast, opponents are accused of 'using politics to stop science' (*Feedstuffs* 1994c) rather than considering a new product solely on its technical merits.

A third discursive claim is that biotechnology is merely an extension of nature and consequently can be expected to be perfectly safe. For example, Monsanto ran an advertisement featuring a blond youngster playing with his dog on a rolling hillside. Crowning this image is a caption which reads 'Without chemicals, life itself would be impossible' (Kleinman and Kloppenburg 1991: 434). Rather than a new form of knowledge fraught with dangers, biotechnology is depicted as a benign, neutral science 'built on concepts and practices handed down through the ages' (Plein 1991: 481).

Finally, Kleinman and Kloppenburg cite as discursive an element which they call the 'hegemony of the free market'. By this they mean that the interests of the biotechnology companies are equated with the national welfare of the country, recalling the much quoted claim by a past chairman of General Motors that 'What's good for General Motors is good for the USA'. Conversely, those who try to slow up the advance of new biotc nologies are branded as hurting the competitive position of the Unitcc States in the world market for biotechnology products. Among others, both the director of the Office of Agricultural Biotechnology and the director of Intergovernmental Affairs at the US Department of Agriculture (USDA) have publicly emphasised the important role to be played by biotechnology in making America more competitive abroad (Plein 1991: 478).

All of these discursive elements can be seen in the campaign to promote bST.

In a fact sheet included in a promotion kit issued immediately after its bST product, Posilec, was approved for sale, Monsanto depicts bST as simply another step in a series of improvements for the dairy industry which began in the 1950s. Furthermore, bST is said to have a valuable contribution to make to the ongoing campaign against world hunger:

> With world food demand expected to double in the next 50 years and developing nations seeking nutritionally adequate diets the gains in greater efficiency realised through supplemental-BST use can help to boost total milk production in areas where more milk is needed.
>
> (*Feedstuffs* 1993: 4)

Supporters of bST technology have consistently cited the extensive volume of scientific data validating the product and its positive review by FDA scientists. 'Science is on our side', one Monsanto spokesperson told the *New York Times* (Schneider 1989a). On another fact sheet Monsanto includes quotes from the FDA, the National Institutes of Health (NIH), the *Journal of the American Medical Association*, CAST (Council of Agricultural Science and Technology) and *Pediatrics* magazine describing the safety to consumers of bST.

To emphasise the 'natural' quality of their product, Monsanto as well as other biotechnological supporters never use the acronyms bST or rBGH, preferring to label it BST. Similarly, the company has launched several lawsuits against dairy companies for making statements implying that their products were free of bST and therefore superior to milk treated with Posilec. One advertisement run jointly with two other chemical/pharmaceutical companies in the farm press in 1989 was entitled 'You've had BST and cookies all your life', an obvious attempt to reassure consumers that BST has always occurred naturally in milk and that the present product is therefore no different, despite the fact that it has been produced more efficiently by the company than by nature itself (Kleinman and Kloppenburg 1991: 443).

Finally, readers are cautioned that if the political battle over milk labelling drags on in the United States, domestic dairy farmers will lose the export market to a number of foreign nations (i.e. Algeria, Brazil, Honduras, Romania, Russia, South Africa) who have already approved bST and who will be able to produce milk more cheaply at home. The United States, it is claimed, faces 'a real penalty in the marketplace competition for not accepting the technology' (*Feedstuffs* 1994a: 8).

The decision over the future of bST has been contested in the courts, at federal regulatory agencies, in legislatures and in the media. In contrast to the other two cases which have been featured here – acid rain and biodiversity loss – there has been no real epistemic community willing to sponsor and legitimate the career of the problem. Similarly, for reasons already cited, the use of bST did not appear on the agenda of UNEP and its international organisational network. It remained primarily an American domestic issue, despite the fact that the world agricultural system is thoroughly globalised.

Although opponents of bST have won some victories in the courts and managed to convince state legislatures in Wisconsin and Minnesota to ban the use of bST until a final decision by the FDA was reached, these were largely symbolic victories which stalled but did not stop its introduction. Probably the most effective tactic in contesting the issue has been the pressure which has been applied to the retail/wholesale grocery store industry.

If any agricultural technology meets significant consumer resistance, grocery chains and other high-volume merchants use their clout as high-volume customers to block the implementation of that technology. This 'gatekeeper' role is so significant that other segments of the food industry (food processors, manufacturers) quickly follow suit (Browne and Hamm 1990: 43). Early on in the campaign against bST, Rifkin and his coalition succeeded in convincing several large chains and dairy cooperatives to announce a bST-free policy. One high-profile supporter has been Ben and Jerry's Ice Cream of Waterbury, Vermont, a firm well known for its support of social causes. In 1989, Ben and Jerry's announced that millions of pints of their product would bear labels urging a halt to the development of bST in order to 'save family farms' (Schneider 1989b: 4). In their recent 'Consumer Warning', the Pure Food Campaign advises readers to patronise those companies that have rejected genetically engineered milk; as well as Ben and Jerry's these include Kroger Supermarkets and 7–Eleven

Stores, the largest chain of convenience stores in the US. The campaign further urges readers to telephone a list of twelve large companies including McDonald's restaurants, A & P supermarkets, Nestlé Carnation and Dannon Yogurt to 'tell them that you will not purchase their products until they provide you with *written assurance* that their products are and will remain free of rBGH'. Perhaps buoyed by the success of environmentalists in pressuring McDonald's to switch from styrofoam containers to paper wrappers, The Pure Food Campaign launched an 'Adopt-a-McDonald's' campaign in 1993 which sought to secure a pledge from the fast food chain that they will not serve dairy products or meat containing bST in the future (Gooderham 1993: A-4).

Another tactic which has become central in contesting the bST issue in the wake of FDA approval in the US has been the labelling of products that contain milk from untreated cows. Correctly sensing that this could have significant consumer impact, the FDA has issued strict guidelines as to what these labels can and cannot say. In particular, it has introduced the 'proper context' principle which stipulates that a statement of non-use be qualified with an accompanying statement about the safety of bST-treated milk or an explanation of why it has not been used (Carlson 1994: 3).

However, opponents have challenged this edict on several fronts. In May 1994, two large Upper Midwest dairy companies – Marigold Foods Inc and Land O' Lakes – introduced a new line of fluid milk which they certified as being bST-free on the basis of signed affidavits from producers. Furthermore, two states (Vermont and Wisconsin) have introduced new anti-bST laws. Vermont's law, which required mandatory bST milk labelling, was immediately challenged in court by the International Dairy Foods Association (IDFA), a trade lobby group. In April 1994 the state of Wisconsin passed a less stringent law than Vermont's which gives processors the option to state on labels that cows were not treated. According to an industry report attributed to the IDFA, dozens of actions are now pending from scattered states, cities and school boards ranging from mandatory labelling to outright refusals to accept products involving bST (*Feedstuffs* 1994b: 11).

Monsanto and the Animal Health Institute, a lobby group which coordinates the public relations in favour of bST, have undertaken their own campaign. In addition to issuing a promotion kit, the company plans to select dairy producers for demonstration projects, offer training sessions for interested farmers and veterinarians and sponsor a cable television show produced by the American Medical Television System (*Feedstuffs* 1993: 4).

Opponents have long held that Monsanto and the other three companies seeking approval for bST have unduly influenced policy at federal agencies responsible for assessing this product. In January 1990, the *New York Times* published an interview with Richard J. Burroughs, a veterinarian who had worked for three years (1985 to 1988) on the FDA review of manufactured bovine somatotropin until he was fired, allegedly for being too critical of the agency's close ties with the Animal Health Institute. Burroughs claimed that the regulatory agency was becoming 'an extension of the dairy industry' and that important flaws in safety studies submitted by the manufacturers had been summarily dismissed, a charge denied by FDA officials (Schneider 1990b: A–21). In a similar fashion, the Consumers Union, the publisher of *Consumer Reports* magazine, has suggested that unpublished documents which indicate health and safety concerns with bST have never been publicly released by the FDA (Andrews 1991: A–18). In 1990, based on documents obtained through the Freedom of Information Act, Jeremy Rifkin initiated a lawsuit in which he contended that the National Dairy Promotion and Research Board had awarded a $1 million contract to a Washington public relations firm to assure the public that bST was safe for use. At the same time, Rifkin petitioned the FDA to stop promotion of the product by agency officials (*New York Times* 1990: A–25). All this contrasts significantly with the British case where the veterinary products committee did not hesitate to reject the application for licences by Monsanto and Eli Lilly on the basis of similar evidence to that available to the FDA.

For the most part, the mass media has not functioned as an agenda builder in the conflict over the future of bST. As Kawar (1989) has noted, after a short period of interest in the initial debate over biohazards during the years 1975 to 1978, media attention declined and shifted to the growth of the biotechnology industry after 1980. While the media have periodically reported on the bST controversy, they have made no effort to frame it as a 'moral panic' (Ungar 1992) in contrast to some other environmental problems such as global warming. For example, the *New York Times* carried fourteen stories on the subject during the period 1989 to 1990 when opposition to bST first peaked. However, editorially, the newspaper has consistently supported its introduction, entitling their opinion pieces 'riskless hormone' (20 September 1990) and 'Hiding behind hormones in milk' (5 May 1989).

To a large extent, this lukewarm media interest has been the result of the failure of claims-makers to present evidence which directly links the ingestion of bST-altered milk to any human disease such as cancer or leukaemia. Furthermore, despite some success with special interest groups such as farmers and animal rightists, biotechnological critics such as Rifkin have so far been unable to spark a national movement against genetic engineering, thereby leaving the field open for other issues such as acid rain, ozone depletion and toxic waste (Sherlock and Kawar 1990: 121). Finally, the regular news sources upon which the mainstream media depend for guidance have not registered undue alarm over the possibility of widespread bST use. The scientific media (*Science, Science News, Nature, Journal of the American Medical Association, New England Journal of Medicine*) have generally been supportive of the biotechnology industry while many prominent environmental organisations have had relatively little to say on the issue. As a result, bST as an environmental issue, like that of genetic engineering in general, has been shuffled aside by most journalists 'to the private sector and the remote halls of regulatory agencies' (Kawar 1989: 736).

CONCLUSION

While it has commanded some measure of attention, the introduction of genetically engineered milk has not yet been legitimated in four key arenas: science, media, government and the public. There are several factors which account for this failure.

Unlike most environmental problems, concern over biotechnology does not originate *within* science but rather exists as a critique of science. As a result, it has lacked credible scientific champions who can build a viable case for banning the use of this synthetic animal hormone. The only oppositional data which does rest on a scientific foundation – the elevated incidence of mastitis in cows treated with bST – is not sufficient to create a sense of widespread public alarm.

This may change, however, as wider, non-experimental usage brings further medical research. In the wake of bST approval, Samuel Epstein,[6] professor of occupational and environmental medicine at the University of Illinois and the leading proponent of the theory that cancers have environmental causes, told the FDA that he had grave concerns about the risk of breast cancer in women who drink bST milk because the hormone increases insulin growth factors which have been identified as a growth factor for human breast cells inducing and maintaining malignancy (Carlson 1994: 12). Any empirical confirmation of Epstein's thesis would certainly introduce a new dimension into the controversy over bST.

Second, unlike biodiversity loss, the introduction of bST has not been framed as a social justice issue on a global scale. While bST opponents have successfully linked it to the decline of the family farm, they have not warranted the problem in terms which would appeal to the international network of agencies, commissions and organisations centred around development issues. Biodiversity loss is also centrally linked to biotechnology but it has been presented as a problem which transcends the interests of one occupational group in one region of the United States. Significantly, some of the key figures in the campaign against biotechnology (Pat Mooney, Cary Fowler, Vandana Shiva) directed their energies into the biodiversity problem rather than against the genetic modification of domestic animals.

Third, the economic benefits of banning bST accrue to social actors with relatively little power – small dairy farmers and cooperatives – rather than to the more powerful multinational chemical and pharmaceutical corporations and large agricultural cooperatives. This contrasts with biodiversity loss where most of the same companies would suffer from the disappearance of rare species of plants and animals which are possible sources of new drugs and chemicals. Even in the case of acid rain, the problem was most successfully promoted in Sweden and Germany where the giant forestry companies, hunting associations, nuclear facilities and other powerful economic actors perceived that they would lose if something were not done about the coal-fired plants which were thought to be causing forest die-back.

Nevertheless, the campaign against bST may yet enjoy some success. On its side are a powerful symbolic rhetoric (the purity of nature as embodied in milk), a strong adversarial quality, and an identifiable economic disincentive for potential users (i.e. the spectre of milk surpluses and price drops).

Much may depend on the reaction of the European Union (EU). In 1989, the European Parliament supported a resolution calling for a worldwide ban on hormones and other substances used to improve productivity in livestock production. In 1994, the EU seemed poised to approve a seven-year ban on bST but, in the final hour, the Commission opted instead for a further one-year moratorium. This was later extended to the end of the century but the EU permitted some field testing. With butter mountains and wine lakes a thing of the past, the EU is not eager to support a technology which might lead to milk surpluses, especially insomuch as milk quotas are in place until the year 2000. Furthermore, as evidenced by the widespread recent support for the campaign against forest practices in the logging industry of British Columbia, the Green lobby in Europe remains a significant force. At the same time, with the milk productivity of European dairy herds declining in relation to their US counterparts, there may be a strong counter-pressure to eventually accept the technology. If, however, the issue of bioengineered animal hormones rises in the agenda of European environmental groups and links up with the current protest action in North America, then an effective international campaign may yet be assembled.

10

ENVIRONMENTAL CONSTRUCTIONISM AND THE POSTMODERN CONDITION

In this final chapter, I will deal with the social construction of environmental risk and knowledge in the context of the larger theoretical debate over modernity versus postmodernity. For the most part, environmental sociologists have eschewed involvement in this debate, preferring to deal with more empirically grounded research problems. However, as we will see, there has recently been inceased activity on this front, notably Beck's development of the concept of 'reflexive modernity' in the context of a 'risk society' and Spaargaren and Mol's attempt to trace the contours of 'ecological modernisation' as a theory of social change.

To a considerable extent the modern–postmodern dichotomy resembles that of a concealed patch of quicksand: initially encouraging but ultimately fraught with peril. As Featherstone (1988: 195) has observed, few other recent academic terms can claim to have enjoyed such popularity as postmodernism and yet still be regarded by many as an 'ephemeral fashion' or as a 'rather shallow and meaningless intellectual fad'.

Despite its checkered career at the cutting edge of social theory, there is a serious issue at stake in conceptualising the postmodern. While it can mean many things from a style of architecture to a method of approaching literature, in sociology it has most usefully been treated as 'an epochal shift or break from modernity involving the emergence of a new social totality with its own distinct organizing principle' (Featherstone 1988: 178). This principle states that the certainty of the modern era, constructed as it was on widely shared and accepted notions about economic progress, has been pulverised, leaving a fragmented, chaotic world which is utterly devoid of meaning. If production and industrial capitalism were the hallmarks of modernity, postmodernity is characterised by a social order which is dominated by 'simulations': artificial representations or copies of real objects or events (Baudrillard 1983). Not surprisingly, the symbol of the

178

present age for many postmodernist scholars is the Florida theme park, 'Disney World', which is said to epitomise the substitution of a sanitised, controlled history and cultural experience for the real thing.

As it is largely a kind of indictment of the 'plastic' nature of contemporary culture, postmodernism really does not say much about the economic and political changes which are involved in the transition from the modern to the postmodern. One exception to this is the Marxist critic Fredric Jameson (1984), who depicts postmodernism as a new socioeconomic stage of capitalism in which capitalist exchange relations have penetrated the spheres of information, knowledge, computerisation and consciousness and experience itself to an unparalleled extent. However, as Kellner (1988: 261) has correctly observed, Jameson's account deliberately reduces postmodernism to a 'moment within the new stage of capitalism' and might better be characterised by some other concept, for example, 'multinational capitalism'.

What has all of this got to do with the social construction of environmental problems? First, the unravelling of modernist discourses concerning economic growth, development, risk and science can be seen as being consistent with postmodernist explanations. As Brian Wynne (1992) has argued, the 'modernist paradigm of a singular, unconditional rationality of which dominant risk discourses are a pillar' has been undercut by the events of the last thirty years. This deconstructive process first began in 1962 with the publication of *Silent Spring*, Rachel Carson's indictment of the agrochemical establishment and its reckless use of pesticides, followed immediately by Barry Commoner's (1963, 1971) critique of post Second World War industrial technologies and their destructive effects on the ecology of the planet. In their wake, a plethora of commentators both from within the walls of academia and from environmental movement organisations systematically unmasked the basic lack of certainty which surrounded the conduct of science and the introduction of new technologies. Nuclear power, plastics, herbicides, toxic waste incinerators: each of these technologies was subsequently demonstrated to have serious flaws which had not been evident when they were first hyped as tools for a brighter future. Furthermore, the process of doing science itself was revealed to be inherently sociological, a 'web of conventions, practices, understandings and "negotiated" indeterminacies' (Grove-White 1993: 22).

At the grassroots level, this eroding faith in science was sparked by a progression of fictional (*The China Syndrome*) and real life (Bhopal, Chernobyl, Love Canal, Three Mile Island) chemical and nuclear disasters in which those in control of technology did not appear to know what they were doing, choosing instead to stonewall the truth. This perception was further boosted by a steady stream of 'pseudoevents' (Boorstin 1964) staged by Greenpeace and other environmental claims-makers which depicted whalers, nuclear operators, forestry companies and others as evil personified. The upshot was the weakening of confidence in science and industry much as the Watergate affair in the United States destroyed people's faith in the conduct of politics and government. That is not to say that economic growth, full employment and technological progress faded uniformly as desiderata but they were now tempered by a lingering feeling that those in charge were not always being straight with the citizenry-at-large. Lash and Wynne (1992: 7) refer to the emergence of a 'private reflexivity' in which people began to express this underlying distrust within the boundaries of their own semi-private worlds and in their own vernacular.

As a consequence, what the postmodernists call the 'grand narratives' or 'metanarratives' (Lyotard 1984) of the past were increasingly becoming delegitimated. Even though many people were required by virtue of job-related dependencies to feign trust in organisations or institutional complexes, this trust and credibility was, in fact, only skin-deep, concealing a deeper level of ambivalence which might be recognised as 'an essential condition of postmodernism' (Wynne 1992: 296). If, as Lash and Urry (1994: 257) observe, postmodernity inherently 'proclaims the end of certainty', then the deconstruction and reconstruction of environmental risks and knowledge are destined to become characteristic of tomorrow's society.

A further echo of postmodernism is suggested by Harries Jones (1993: 5) who points out that, inspired by the success of Greenpeace, environmental movement organisations have increasingly framed their 'knowledge interests' primarily through media images. Harries Jones terms this 'iconic praxis' and observes that these images have come to be artificially constructed. In particular, he argues that the 'eco-dramas' staged by Greenpeace have gone from being real to being simulated in order to conserve organisational resources and to protect the safety of activist members. Thus, following Baudrillard (1983: 2), environmental advocacy is said to have entered the regions of the 'hyperreal' where nothing is as it seems and pseudo-events such as the Pure Food Campaign's 'Adopt-a-McDonald's' initiative (see Chapter 9) reign supreme. It is worth noting that the staging of pseudo events and the use of simulational tactics are not entirely unique to the environmental movement organisations of the present era. In fact, in the 1920s, a fierce controversy raged for a while in the American conservation movement over the tactics of 'nature fakers' – nature writers such as Ernest Thompson Seton and Charles G. D. Roberts who attributed human personality characteristics and abilities to wild animals (Lutts 1990; Schmitt 1990). However, the more sophisticated and pervasive manipulation which can be created today through the use of digital computer technologies leads to the spread of a form of 'digital escapism' characterised by the emergence of a 'virtual conscience': if we can project ourselves into a perfect world with a few strokes of the keyboard, why worry about the state of the environment which we have left behind in the real world (Cormier 1994)?

Another commentator who links postmodernism and the construction of environmental problems via the process of icon construction is Andrew Szasz (1994). Szasz begins his discussion with much the same observation as Harries Jones: that environmental advocates have mastered the fine art of icon-making and have used it to create environmental issue stories that feature 'repetitive, highly stereotyped, frightening imagery'. News consumers form their attitudes towards toxic waste, nuclear power, the greenhouse effect and other environmental topics almost totally on the basis of these superficial images rather than on a more sustained cognitive assessment of non-visual information. While you would think that the episodic attention typical of a postmodern society would automatically lead to a rapid decline in interest as the news moves to other stories, this is not necessarily true. Rather, Szasz proposes a model wherein postmodern issue creation is transformed into more traditional forms of social action, notably the formation of grassroots citizens' groups. This is most likely to occur when media images connect with personal experience, 'making the icon the perceptual basis for a more traditional politics of the social movement' (1994: 83). This, Szasz argues, is what happened in the early 1980s, in the case of hazardous waste, when a battery of television-mediated images of toxic waste leaking from broken fifty-five-gallon drums, clean-up crews encased in protective safety gear and boarded-up homes dovetailed with real-life feelings of fear and anger in neighbourhoods impacted by toxic facilities.

Despite such currents of postmodernism, most environmental researchers who have considered the matter have shied away from adopting a postmodernist perspective. As Wynne (1992: 296) explains, postmodernist theory in its stronger forms (e.g. Lyotard) projects an image of the new society as being incoherent, a Babel of multiple social identities. Instead, they have chosen to follow the lead of Giddens (1990, 1991) and opt for a revised or updated form of modernism.

Perhaps the best-known example of this middle way is Ulrich Beck's theory of the 'risk society'. Beck's thesis starts with the premise that Western nations have moved from an 'industrial' or 'class' society in which the central issue is how socially produced wealth can be distributed in a socially unequal way while at the same time minimising negative side effects (poverty, hunger) to the paradigm of a 'risk society' in which the risks and hazards produced as part of modernisation, notably pollution, must be prevented, minimised, dramatised or channelled. Both the former 'wealth distributing society' and the emergent 'risk distributing society' contain inequalities and these overlap in areas such as the industrial centres of the Third World.

One important feature of the risk society is the way in which the past monopoly of the sciences on rationality has been broken. Paradoxically, science becomes 'more and more *necessary*, but at the same time, *less and less sufficient* for the socially binding definition of truth' (Beck 1992: 156). Beck contrasts the rigid 'scientific rationality' which is rooted in a critique of progress with a new 'social rationality' which is rooted in a critique of progress. Under pressure from an increasingly edgy public, new forms of 'alternative' and 'advocacy' science come into being and force an internal critique. This 'scientisation of protest against science' produces a fresh variety of new public-oriented scientific experts who pioneer new fields of activity and application (e.g. conservation biology). In a similar fashion, monopolies on political action are said to be coming apart, thus opening up political decision-making to the process of collective action. One example of this is the entry of the 'Greens' into parliamentary politics in Germany in the 1980s.

Finally, the dynamic of reflexive modernisation leads to greater indi-vidualisation. Unbound from the strictures of traditional, pre-modern societies, the new urban citizens of the industrial revolution were supposed to reach new levels of creativity and self-actualisation. However, this did not happen, largely because a new constraint – the 'culture of scientism' – invaded every part of our lives from risk construction to sexual behaviour. Now there is a chance for the individual to once again break free and choose their own lifestyles, subcultures, social ties and identities (p. 131). Yet ironically, just as this individualised private existence finally becomes possible, we are confronted with risk conflicts which by their origin and design resist any individual treatment. Beck does not use the phrase but it is clear that 'global environmental problems' such as the greenhouse effect and the thinning of the ozone layer are key illustrations of this. Thus the 'reflexive scientisation' in which scientific decision-making, especially that related to risk, is opened up to social rationality is vital to the reclamation of individual autonomy.

A second environmental theory of late modernity is the theory of 'ecological modernisation' developed by the Dutch sociologists Gert

Spaargaren and Arthur Mol. Spaargaren and Mol acknowledge several contributions of the 'reflexive modernisation' school (Beck, Giddens, Wynne): their recognition that contemporary global risks have lost their delimitations in time and space, their emphasis on the changing relationship of lay actors and expert systems and their perception that scientists in the era of late modernity can no longer ensure any certainties with regard to environmental risks but rather must share their doubts with the public (Spaargaren and Mol 1992b). At the same time, they criticise the reflexive modernisation approach on the grounds that it is unduly pessimistic. Beck, it is claimed, makes the error of exclusively choosing 'high consequence' risks (e.g. nuclear power) and broadening these to a whole range of environmental issues. By contrast, their own conceptualisation of ecological modernisation as a theory of social change is one of qualified optimism.

By ecological modernisation, Spaargaren and Mol mean an ecological switch of the industrialisation process in a direction that takes into account the maintenance of the existing sustenance base (1992a: 334). Cast in the spirit of the Brundtland Report, ecological modernisation, like sustainable development, 'indicates the possibility of overcoming the environmental crisis without leaving the path of modernization'. Their model is based on the work of the German writer, Huber (1982, 1985), who analyses ecological modernisation as a historical phase of modern society. In Huber's scheme, industrial society develops in three phases: (1) the industrial breakthrough; (2) the construction of industrial society; and (3) the ecological switchover of the industrial system through the process of 'superindustrialisation'. What makes this latter phase possible is a new technology: the invention and diffusion of microchip technology.

Ecological modernisation rejects the Schumacher (1974)-inspired 'small is beautiful' ideology in favour of a large-scale restructuring of production–consumption cycles to be accomplished through the use of new, sophisticated, clean technologies (Spaargaren and Mol 1992a: 340). Unlike sustainable development, there is no attempt to address problems of the less developed countries of the Third World. Rather, the theory focuses on the economies of Western European nations which are to be 'ecologised' through the substitution of microelectronics, gene technology and other 'clean' production processes for the older, 'end-of-pipe' technologies associated with the chemical and manufacturing industries. In contrast to Schnaiberg's 'treadmill of production' perspective, capitalist relations of production, operating as a treadmill in the ongoing process of economic growth, are treated as largely irrelevant (Spaargaren and Mol 1992a: 340–1).

According to Udo Simonis (1989), a German environmental policy analyst, the ecological modernisation of industrial society contains three main strategic elements: a far-reaching conversion of the economy to harmonise it with ecological principles, a reorientation of environmental policy to the 'prevention principle' (seeking a better balance between stopping pollution before it happens and cleaning it up later on) and an ecological reorientation of environmental policy, especially by substituting statistical probability for 'prove-beyond-a-doubt' causality in legal suits against polluters. Unfortunately, little is said about the social and political barriers which are likely to be faced in trying to implement these strategies, especially in countries other than Germany and the Netherlands where the environment is already a major priority.

While both of these theoretical attempts to rechart the frontiers of late modernity are commendable insomuch as they bravely attempt to relate the 'environmental crisis' to the character of modern society, they nevertheless fall short on several counts.

As Lidskog (1993) has pointed out in his review of *Risk Society*, Beck contradicts himself by arguing that the planet is in increasing peril due to an escalation of objectively certifiable global risks and, at the same time, insisting that risks are entirely socially constructed and therefore do not exist beyond our perception of them. This reflects a long-standing tension in environmental sociology as a whole between the role of the sociological analyst and that of the environmental activist. Catton and Dunlap's HEP/NEP dichotomy is the epitome of this but it runs through much of the rest of the literature as well, surfacing most recently in the 'realist' approach of Benton, Dickens, Martell and other British sociological thinkers who seek to put nature back into the nature–society relationship.

Ecological modernisation theory, by contrast, is hobbled by an unflappable sense of technological optimism. All that is needed, they suggest, is to fast forward from the polluting industrial society of the past to the new superindustrialised era of the future. Yet the silicon chip revolution which is the basis of this superindustrialisation is by no means environmentally neutral as the theory of ecological modernisation suggests (see Mahon 1985). Furthermore, it is worth remembering that nuclear power was also touted as a 'clean' technology until its more undesirable features became known.

As a sociological explanation, the theory of ecological modernisation is as much prescriptive as analytic. Spaargaren and Mol, in fact, say little about the power relations which characterise environmental processes, assuming somehow that good sense will automatically triumph. Yet, as Gould *et al.* (1993: 231) have argued, sustainability, the guiding concept behind ecological modernisation, is as much a political–economic dimension as an ecological one: what can be sustained is only what political and social forces in a particular historical alignment define as acceptable. Recognition of this is far more evident in Beck's concept of a risk-distributing society than in the ecological modernisation which Spaargaren and Mol see as rapidly approaching. To be fair, it should be noted that Spaargaren and Mol are not entirely oblivious to this weakness. Indeed, they qualify their theory of ecological modernisation by observing that it is 'limited insofar as it deals with only the industrial dimension of modernity, neglecting dimensions of capitalism and surveillance, and because it narrows the concept of nature to the sustenance base' (Spaargaren and Mol 1992a: 341).

The social constructionist approach which I have adopted in this book follows Ian Welsh's (1992) characterisation of the environment as 'a site of intersecting and competing social and cultural definitions and interests'. Contested are the nature and gravity of environmental threats, the dynamics underlying them, the priority accorded one issue versus another and the optimal means for mitigating or ameliorating conditions which have come to be defined as problematic. The parties involved in the contestations include private industry, government, regulators, scientists, environmental groups, community organisations, trade and professional groups and, increasingly, grassroots 'victims'. What is ultimately most significant here is the process through which environmental claims-makers influence those who hold the reins of power to recognise definitions of environmental problems, to implement them and to accept responsibility for their solution (Hannigan 1995).

This conceptualisation of environment and society fits easily into the postmodernist image of a world which is 'contingent, hazardous and erratic' (Bauman 1994: 143). In contrast to several other social scientific approaches to environmentalism and the environment, it does not contain implicitly or explicitly any kind of evolutionary component. For example, those commentators who use as their launching point the contrast between the dominant and the new ecological paradigms, suggest that it may take a long time but eventually the latter will justly triumph. Lester Milbrath's (1989) book on social learning and the environment is a leading example of this. Similarly, even though the basis of its optimism is quite different, ecological modernisation theory also possesses an evolutionary flavour in its prediction of an 'ecological switchover' to a new superindustrial, sustainable society. A social constructionist perspective does not preclude the arrival of an ecologically kinder and gentler society, perhaps even via Beck's 'reflexive modernisation' process, but it cautions that environmental issues and problems constantly rise and fall as do our definitions and understandings of nature, ecology, risk and other elements of the environment–society nexus.

There is one final way in which environmental constructionism and the postmodern condition interweave which has not been explicitly recognised by past sociological commentators.

In discussing the nature of postmodern urban landscapes, Sharon Zukin (1988: 229–30) identifies a process of 'cultural appropriation', whereby a rump of new middle-class 'gentrifiers' in transitional neighbourhoods assert an opposing claim to this space, a claim based not on occupancy or entitlement but on appreciation of the space (or its built form) as a product of cultural consumption. Zukin uses the example of Clerkenwell near London's Smithfield Market which was transformed from a little-known working-class and commercial neighbourhood with scattered sites of historical interest into an upscale area with historical walking tours, architects' and designers' offices and other elements of a new, reorganised postmodern cultural landscape.

It seems to me that this process of cultural appropriation has been equally characteristic of environmental claims-making for quite some time. As part of the back-to-nature movement of turn-of-the-century America, city-dwellers redefined the basic nature of the countryside in aesthetic rather than economic terms, even going so far as to denigrate local farmers as simpletons who were unable to grasp the real meaning of nature. Similarly, today, environmental activists register moral claims to a wide variety of natural spaces from vacant lots in urban neighbourhoods to the rainforests of tropical nations. Like gentrifiers fascinated by historic landscapes, these claims-makers justify their actions on the basis of some superior intellectual insight; for example, that derived from a knowledge of ecosystems. Such claims are frequently opposed not just by rapacious corporate polluters and intransigent bureaucrats but by ordinary locals who frame their spaces in contradictory terms. Thus ecologically minded city-dwellers who permit their front yards to become a wilderness garden clash with neighbours who prefer a tidy lawn and fear a decline in their property values. Burgess and Harrison (1993) describe how residents of Rainham, Essex, could not relate to environmentalists' claims that a local marsh was put in danger by a proposed theme park because they had always regarded it as wasteland. On a grander scale, indigenous peoples in the rainforests of the South often do not share the perceptions of professional environmentalists that large tracts of rainforest must be reclaimed and protected by turning them into wilderness preserves. In such cases, the appropriation of natural environments involves a clash between opposing cultural constructions, one rooted in a vernacular, the other in a new ecological sensibility.

CONCLUSION

In opting for a social constructionist approach, I have conceptualised the environment neither as an economic resource to be exploited, conserved or transformed, nor as a distinct social actor in its own right engaged in a series of dynamic relationships with humans. Rather, I have viewed it as the site for a repertoire of definitional and contestatory activities, many of which are increasingly taking place in a global context. Consistent with this, I have argued that the core of a new environmental sociology should principally lie not in documenting the social distribution of environmental value clusters nor in fleshing out a 'new human ecology' for the 1990s but rather in understanding how claims about environmental conditions are assembled, presented and contested.

As it has become more visible in the sociological study of the environment, the social constructionist perspective has begun to attract a sharp volley of critical fire. Some criticism has come from the 'new human ecologists' (see Dunlap 1994) but the central source has been the 'realism' school of environmental sociology in Britain (Ted Benton, Peter Dickens, Luke Martell).

In recent works, both Benton (1994) and Martell (1994) seize upon the same examples as evidence that the relativist or social constructionist approach to the environment is 'oversocialised'.

First of all, both writers cite a quote from Kitsuse and Spector (1981) which is found in Steven Yearley's book, *The Green Case* (1992). This states that the social problems analyst should suspend any interest in the 'objective reality' of a social problem in favour of examining how claims are constructed.

Second, both Benton and Martell attack the 'extreme relativism' of social constructionism by focusing on Keith Tester's (1991) historical sociology of animal rights. Benton (1994: 45) is amused by Tester's argument that a fish is only a fish when socially classified as such, noting that such comments imply that 'at most the notion of a reality external to discourse is acknowledged as an unknowable ghostly presence'. Martell also cites the example of the fish but goes further, arguing that Tester's claim that nature lacks any objective empirical causal powers is 'implausible'. In addition, he alleges that Tester errs by denying the manifest ideological tenet of animal rightists (i.e. animals have rights because they are of equal value to humans) in favour of a latent social psychological function: concern with animals helps us to define ourselves as 'superior, moral beings' (1994: 132).

Unfortunately, these criticisms paint a rather extreme and absolute portrait of the social constructionist perspective. While it is true that some strict constructionists probably go too far in focusing exclusively on the interpretations and practices of participants in social problems construction, contextual constructionists such as Best and Rafter actively encourage the use of empirical data in the evaluation of claims where this is deemed appropriate.

Similarly, it is misleading to tar all those who adopt a social constructionist perspective on the environment with the brush of absolute relativism. For example, the assertion that global warming should not necessarily be taken at face value as an established scientific fact but rather be seen as something which is open to the social construction of scientific and popular knowledge does not constitute a denial that greenhouse gas emissions exist or that they might possibly have global impacts. Rather, what is being suggested is that the 'natural effects' cited by Martell (i.e. resources, waste-absorbent and food production capacities, climatic and atmospheric effects) may be visualised in a number of quite different, even contradictory, ways and that in turn these interpretations vary according to a variety of factors: interests, cultural background, etc. This is even more likely to be the case where these effects are not directly experienced but are knowable only in the form of scientific data.

Even so, as Buttel and Taylor (1992, 1994) have observed, it is doubtful whether most environmental sociologists are particularly well qualified to evaluate the veracity of environmental claims, especially those which are global in scope. Nor are they in the best position to 'make their science useful by suggesting the alternatives that will accomplish the [policy] objectives' as Borgatta (1994: 41) has advocated. With little formal training in the environmental sciences, they are often inclined to work from popularised accounts of science which are not always very accurate. Furthermore, Buttel and Taylor allege that environmental sociologists are far more inclined to relativise or demystify knowledge claims which come from anti-environmental quarters such as industrial corporations or industrially funded scientists than they are those which come from within the environmental movement itself.

More positively, the case for a constructionist framework has been aggressively stated in a recent article in the journal, *Rural Sociology*, by Thomas Greider and Lorraine Garkovitch (1994). Greider and Garkovitch argue that the role of the environmental sociologist should lie not in a quest for some elusive new model which causally links ecosystem breakdown with social variables (see Catton 1994) but in a return to classic sociological questions of perception and power. In this context, biophysical changes in the environment are meaningful only insofar as cultural groups affected by these changes come to acknowledge them through a self-redefinition. For example, in addressing the political conflict in the American Northwest over the spotted owl, the key question for the sociological analysts should not be the number of owls but the way in which the fluctuating power of the different social actors or claims-makers – loggers, rural businesses, international logging companies, environmentalists – shapes the definition of the situation (Greider and Garkovitch 1994: 21). Greider and Garkovitch conceptualise the idea of global environmental change as a type of 'landscape' and insist that by looking at how this landscape is symbolically created and contested, researchers are both 'incorporating and engaging' (Shove 1994). By doing so, they are contributing to the furtherance of a well-established school of thought in sociology and helping to forge a role for the discipline in the debates over environmental issues.

Critics of social constructionism have generally overlooked its expanding portfolio of successes. As we have seen in Chapter 5, the notion that technological and environmental risks are socially constructed has become central to contemporary analyses of risk perception and management such as those presented by Jasanoff (1986, 1990). As discussed in Chapter 3, the media's role in socially constructing environmental issues and problems has been widely recognised. Similarly, the paradigm of the social construction of science has begun to be fruitfully extended to the environmental field. Finally, the moral entrepreneurship of environmental movement organisations such as Greenpeace and Friends of the Earth has been extensively documented.

In this book, I have attempted to bring these elements together in a more systematic model which focuses on the key tasks and players in the social construction of environmental problems. In doing so, I have sought to demonstrate that constructing and implementing knowledge claims about the environment is far from a simple, straightforward task. Different problems arise in different ways in different contexts. Some never fly despite a seemingly solid basis for concern. Others create a large splash in the media but sink in the quagmire of political decision-making. Rather than trying to capture this in the form of a cyclical model such as that proposed by Downs (1972), I have depicted the progress of putative environmental problems much like Tamino's trials in Mozart's opera, *The Magic Flute*. Thus potentially successful environmental claims must undergo trials by science, public opinion and politics. At the risk of mixing metaphors, it is similarly helpful to visualise environmental claims-making as a version of the classic 'Snakes and Ladders' game where it is possible to fall back or leap ahead depending on a variety of factors: shifts in the biophysical environment (e.g. cooler or warmer summers), a crisis situation (e.g. Bhopal, Chernobyl, Three Mile Island), a political shift (e.g. Reaganism), issue fatigue (especially with the mass media), a new breakthrough in research, a bestselling book or popular film, or a new presentation 'frame'.

Future research on the environment from a social constructionist perspective would significantly prosper by incorporating a more explicit emphasis on power relations. This synergy has been recognised by several prominent environmental sociologists, notably Allan Schnaiberg and Fred Buttel, but it has rarely been explored in sufficient depth. One encouraging lead is the recent work which has been published on the environmental justice movement, notably that of Stella Capek (1993), which casts it in a framework of contested definitions, frames and meanings. Also of value is a growing body of research on social power and global environmental politics and policy formation (see especially Lipschutz and Conca 1993).

Another under-researched aspect of environmental constructionism concerns the rhetoric and symbolic language used in the assembling and presentation of claims. While there is a danger that constructionist enquiry might eventually become preoccupied with this to the exclusion of other aspects of claims-making, rhetorical devices and strategies are important none the less in linking mechanisms between the different arenas through which environmental claims circulate. Such an approach should emphasise not only the word-centred production of meaning but also image-dominated visual representations (Szasz 1994: 57).

Finally, consistent with the theme of Chapter 10, we need to more fully conceptualise the construction of environmental risks and knowledge within the wider context of sociological debate over modernisation, postmodernisation and change. It is important, for example, to gauge whether the social construction of risk is central to late modernity as envisioned by Beck and Giddens or more characteristic of the fragmented, uncertain, image-dominated world described by postmodern writers such as Baudrillard and Jameson.

In pondering this theoretical issue, it is worth noting some recent comments by the British sociologists Scott Lash and John Urry. Lash and Urry (1994: 296–9) argue that central to recent changes in the social construction of nature have been transformations in the structural import-ance and character of consumption. While the development of 'an overwhelming global consumerism' has had profoundly negative implica-tions for the physical environment, it has also, paradoxically, turned nature protection itself into a consumer activity which is culturally constructed.

This fusion of consumerism and nature protection has been linked to the travel and other consumption patterns of the 'new middle class'. For example, Munt (1994) visualises an 'oppositional postmodern travel' in which ecological treks and cruises to exclusive, restricted Third World destinations such as the Galapagos Islands and the Costa Rican rainforest satisfy their craving for social and spatial distinction from the 'golden hordes' who populate mass tourism. Environmental sensitivity thus emerges as a symbolic badge of a progressive middle-class lifestyle bestow-ing both status and collective identity.

Lash and Urry (1994: 297) observe that an important element of this new consumer style is a 'heightened reflexivity' about places and environ-ments. In particular, demands for consumer rights are extended from the human to the natural world. In doing so, the environment is constructed in a significantly different manner from that which emphasises straightfor-ward exchange values.

This consumption-related perspective illustrates a potentially fruitful route which environmental sociologists might undertake in the future in order to incorporate a social constructionist approach into a more general analysis of social structure and social change. One major benefit of such an analysis would be to synthesise the political economy and social construc-tionist approaches in an innovative manner in which the emphasis would be shifted from production to consumption as the central focus of the nexus between the economy and the natural environment.

NOTES

1 ENVIRONMENTAL SOCIOLOGY: ISSUES AND THEORETICAL APPROACHES

1 Addressing this point directly, Sorokin comments, 'A reader of these lines may think Dr Huntington has at his disposal there the detailed record of the Meteorological Bureau of Ancient Rome' (1964 [1928]:191).

2 As it happens, NEP originally stood for New Environmental Paradigm, but Catton and Dunlap renamed it in 1980 in recognition of the increasingly ecological perspective involved in most environmental research (Freudenburg and Gramling 1989: 445).

3 This model was first briefly introduced in Dunlap and Catton (1983) and was elaborated in several papers presented at scholarly meetings in the late 1980s (Catton and Dunlap 1989, 1989). See Dunlap (1993: 734–5).

4 The phrase 'reflection hypothesis' was first used by Gaye Tuchman (1978) to describe the correspondence between what we see on television and what we observe in real life. Tuchman suggested that the 'null hypothesis' was true here; that is, that the media did *not* reflect actual values and behaviours in contemporary society.

5 This hidden politics of environmental agenda-setting is demonstrated in M.A. Crenson's (1972) classic comparative study of the time disparity in establishing air pollution legislation between two midwestern American cities with similar levels of air quality.

2 SOCIAL CONSTRUCTION OF ENVIRONMENTAL PROBLEMS

1 Ibarra and Kitsuse (1993) also outline a set of 'counterrhetorical strategies' which are meant to block claimants' attempts to construct a problem and/or demand action.

2 A recent and notable example of this is Riley Dunlap's chapter entitled 'From environmental problems to ecological problems' in Calhoun and Ritzer (1993). However, as discussed in Chapter 1, Dunlap's analysis is rooted in a normative 'ecological' approach to environmental problems.

3 This includes the 'sociology of risk' which significantly overlaps but is not

coterminous with environmental sociology (see Freudenburg and Pastor 1992; Covello and Johnson 1987; Short 1992).

4 This was suggested at the public hearings on the proposed Alberta-Pacific bleached Kraft pulp mill in Northern Alberta by Cindy Giday from the Northwest Territories who was the lone native (and female) on the Alpac EIA Review Board (see Chapter 5).

5 Total membership of the twelve or so major national environmental organisations in the US increased from about four million in 1981 to roughly seven million in 1988 (Bramble and Parker 1992: 317).

6 Note, however, that in the course of fundraising and lobbying, major conservation organisations are inclined to draw from both rhetorics, anchoring their appeals in both commercial and moral rationales (Yearley 1992: 26).

7 Unfortunately, the *sao la* has recently become threatened with extinction as collectors from around the world attempt to obtain one, even reputedly offering a bounty of up to $1 million (Shenon 1994).

8 This controversy is reflected in the different acronyms that are used in discussing this bioengineered hormone. Opponents attempt to distinguish it from its naturally occurring form by calling it bST or rBGH, while industry and regulatory defenders who claim that there is no difference use the acronyms BST or BGH.

3 NEWS MEDIA AND ENVIRONMENTAL COMMUNICATION

1 For example, Murch (1971) reports that 73 per cent of Americans get their news about the environment from television and 62 per cent from newspapers versus 21 per cent from friends and 12 per cent from other sources.

2 In this chapter, the focus is on media reporting by the everyday press. It should also be noted that there is an extensive array of feature magazine articles and documentary television programmes on nature and the environment. For example, in the United States, the Public Broadcasting Service (PBS) runs a number of popular natural history shows such as *Nature, Nova* and *America's National Parks* as well as operating *Operation Earth*, an ongoing outreach project to educate viewers about the environment. And Atlanta's Turner Broadcasting System (TBS) has its own environmental unit which, among other things, produces *Network Earth*, a weekly magazine show which also appears on the Cable News Network (CNN) under the title *Earth Matters* (Motavelli 1995).

3 The phrase 'Spaceship Earth' was evidently coined by the British economist Barbara Ward as the title of a book she published in 1966 on the links between economics and the environment (Pearce 1991: 11).

4 The only exception to this was the *New York Times* coverage which continued to separate various aspects of environmental issues.

5 Reporters' first choice here is usually a government spokesperson rather than a scientific expert. Sandman *et al.* (1987) suggest that one major reason for this is that reporters generally want two very specific types of environmental risk information: how much of the hazardous substance is in the air and how much of this substance it takes to cause problems.

6 Nearly twenty years earlier, an American researcher (Witt 1974) noted a similar

diversity of environmental sources. Witt's results indicated that the primary news sources of environmental reporters were conservation clubs and organisations and conservation agencies followed closely by business and industry sources. It is worth noting that unlike Cottle, Witt did not extract his sources from media content alone, relying instead on a national questionnaire survey of environmental reporters working for US newspapers.

7 Einseidel and Coughlan (1993) found some revealing differences when they compared the environmental content in Canadian daily newspapers with full-time environmental writers and with those which utilised general reporters. On the whole, there were more environmental stories in the former; the environmental beat reporters were more likely to write longer, more analytical, self-initiated pieces and they were more likely to challenge conventional institutional wisdom.

8 According to a survey carried out by *Editor & Publisher* in the summer of 1970, there were 107 environmental reporters working in the American media, mainly on daily newspapers (Schoenfeld 1980: 456).

9 I witnessed this firsthand when doing participant observation in the newsroom of a national television network in Canada. One day, a senior producer was visibly upset when he received a letter from a viewer charging that the national news broadcast had been giving far too much time to an anti-nuclear protest, despite the newsworthiness of the issue (see Hannigan 1985).

10 There are, of course, some notable exceptions. For example, Lydia Dotto, a veteran Canadian science journalist who has authored books on the space programme, acid rain and global warming, recently wrote an insightful newspaper article on scientific uncertainty and climate change in which she notes that most scientists are not so much accumulating a body of 'facts' as trying to reduce uncertainty, a particularly tough challenge when dealing with the workings of such complex, interconnected systems as the Earth's atmosphere and oceans (Dotto 1993).

11 'Monkey wrenching' or 'ecotage' refers to a wide range of actions taken by environmental activists to disrupt and halt damage to the environment including pouring abrasives into the crankcases of road-building vehicles, pulling up surveyors' stakes and 'spiking' trees by driving long metal spikes into them. The name comes from Edward Abbey's 1976 novel, *The Monkey Wrench Gang*, in which a group of environmentalists plot to blow up the Glen Canyon Dam (see Franck and Brownstone 1992: 190; Manes 1990: 8–9).

4   SCIENCE AS AN ENVIRONMENTAL CLAIMS-MAKING ACTIVITY

1 An exception to this is Germany, where the precautionary principle has been enshrined historically.

2 Scientific concern over pesticide poisoning began more than two decades prior to the publication of *Silent Spring*. As far back as 1945, Rachel Carson herself evidently attempted unsuccessfully to interest *Reader's Digest* in commissioning an article from her on the research being conducted by colleagues at the Paxutent Wildlife Research Center which indicated that the pesticide DDT had adverse effects on the reproduction and survival of birds after repeated

applications (Lear 1993: 33). In the early 1950s, an emerging consensus in the US public health field that the use of chemicals in food production needed to be more strictly regulated led to forty-six days of Congressional hearings. However, the issue was seen as narrow and technical and received little media attention. Unlike the eventual environmental campaign sparked by Carson's book, evidence that pesticides might cause harm somewhere down the road was not as compelling to the media and the mass public as dramatic images of dead birds (Bosso 1987: 80).

In the years 1957 to 1959 there were a series of pesticide-related accidents, notably massive fish mortality throughout New York State due to a gypsy moth spray campaign, and the Great Cranberry Scare, in which cranberry sales fell by two-thirds after some of the fruit was found to be contaminated by residues of the herbicide aminotriazole. Yet these controversies were seen as being isolated and were not sufficient to change the *status quo*.

5 CONSTRUCTING ENVIRONMENTAL RISKS

1 Other notable members of this circle include Steve Rayner and Michael Thompson.
2 Until recently, human sewage from many households mixed together with storm-water in the same pipe. There has since been a vigorous sewage separation programme but some residences still discharge sewage into the storm-water system.
3 One exception to this is a 1984 decision in the *Ferebee v. Chevron Chemical Co* case in the United States which allowed the jury to rely on the testimony of individual physicians in the absence of iron-clad epidemiological evidence concerning injury by exposure to pesticides (see Cronor 1993).
4 As it happens, the Review Board recommended that the mill should not be built unless further studies indicated that it would not pose a serious hazard to life in the river and for downstream users along the Peace–Athabasca river system. Nine months after it agreed to abide by these findings, the Alberta government overturned its own decision and decided to allow Alpac to proceed.

6 NATURE, ECOLOGY AND ENVIRONMENTALISM: CONSTRUCTING ENVIRONMENTAL KNOWLEDGE

1 Grant is one of the more controversial figures in the early wilderness protection movement. A patrician lawyer with close links to many elite figures in business and politics including Teddy Roosevelt, he was among other things a founder of the Save the Redwoods League, the New York Zoological Society and the Boone and Crockett Club. At the same time, he has been called by historian John Higham (1963) 'intellectually the most important nativist in recent American history'. Grant's book, *The Passing of the Great Race* (1921), was for a while a popular selling exposition on the principles of eugenics although it was less successful in subsequent printings. Grant's concern with the subject of eugenics and racial exclusion was shared by a number of other leading

wilderness protectionists of the day including William Hornaday, Fairfield Osborn and Vernon Kellogg.

2 Anna Sewell's book *Black Beauty* was originally published in England in 1877 where it sold more than 90,000 copies. It was republished by the American Humane Education Society in 1890. While designed to increase support for the animal welfare movement, the book also helped to establish a climate for the wider support of wildlife conservation (Lutts 1990: 22–3).

3 In the 1930s, due largely to the efforts of Charles Adams, director of the New York State Museum, and Paul Sears, a plant ecologist, The Ecological Society in the US did make some attempt to bring social scientists and ecologists together in a common forum, notably in a joint symposium of the Society with the American Association for the Advancement of Science entitled 'On the relation of ecology to human welfare – the human situation'. Sadly, two of the leading theorists from the Chicago School, Ernest Burgess and Roderick McKenzie, were unable to attend, leaving August Hollingshead as the only representative of sociology (Cittadino 1993).

4 Kwa (1993: 248) dates the beginning of 'ecosystem ecology' to 1953 when Eugene Odum, a University of Georgia zoologist, published his influential text, *Fundamentals of Ecology*. Soon after, Odum began a series of radioecological studies under the sponsorship of the Atomic Energy Commission (AEC) to determine the impact on the environment of a new atomic weapons plant which was to be built on the Savannah River in South Carolina.

5 The term 'organisational weapon' was first introduced in Philip Selznick's classic (1960) study of the American Communist Party. Eyerman and Jamison (1989) borrow the concept to describe Greenpeace's use of flamboyant and sometimes illegal media-capturing actions to pressure governments and business. Organisations are weapons in such cases when they act in a manner that is considered unacceptable by the community as a legitimate mode of action.

6 In a 1992 interview, Lois Gibbs, the heroine of the Love Canal story, told environmental activist and author Robert Gottlieb: 'Calling our movement an environmental movement would inhibit our organizing and undercut our claim that we are about protecting people, not birds and bees' (Gottlieb 1993: 318).

7 In an article published posthumously, Chavez (1993: 166–7) charges that corporate growers in California have effectively sidestepped many of the provisions of these contracts including those governing the use of pesticides. Chavez observes that many of these same growers were the largest financial contributors in the campaign to defeat Proposition 128 (nicknamed 'Big Green'), a 1990 ballot initiative supported by environmental groups and the UFW which, among other things, would have 'protected California's last strands of privately held redwoods and banned cancer-causing pesticides'.

8 See, for example, Greenpeace (1991).

9 The term 'environmental racism' was evidently coined by the Reverend Benjamin Chavis, former head of the United Church of Christ Commission on Racial Justice and now executive director of the National Association for the Advancement of Colored People (NAACP), a major civil rights organisation in the United States (Higgins 1993: 287).

10 The impetus for this study was a request from Walter Fauntroy, a Congressional representative from Washington, DC, and an active participant in a struggle

in Warren County, North Carolina, to stop the establishment of a toxic landfill containing PCB-laced soil (Bryant and Mohai 1992: 2).

11 It should be noted that the funding for this conference was gold-plated, including among other sources the Ford Foundation and the Rockefeller Family & Associates (Mayer 1992).

12 The Caribbean Ecology Forum (now known as the Karibbean Ecology Trust) is a partial exception, insomuch as it started in the late 1980s as an organisation whose main goal was to halt the toxic trade between Guyana and the United States. Its function has since shifted to providing environmental education to minority youths (Taylor 1993: 272).

7 ACID RAIN: FROM SCIENTIFIC CURIOSITY TO PUBLIC CONTROVERSY

1 In the federal election of March 1983, the victorious Christian Democrats and their allies, the Free Democrats and Christian Socialists, became the champions of saving the forests. There were evidently two reasons for this. First, Christian Socialist political strength was centred in Bavaria where the *waldsterben* (forest die-back) was found to be most extensive. Second, as the chief proponents of nuclear power, the Christian Democrats saw several distinct strategic advantages in pushing a policy of more stringent controls on coal-fired sulphur dioxide emissions. By closing some older power stations and raising the cost of the electricity generated by the remaining utilities, the economics of nuclear power would be enhanced at the same time as political points would be gained by saving the forests. Furthermore, the Greens would be pushed into a policy dilemma since they could not demand both the closing of nuclear power stations and stringent controls on existing coal-fired plants without appearing to be economic 'luddites' (see Boehmer-Christiansen and Skea 1991: 203).

2 There are a number of reasons which have been suggested for the Tory reluctance to embrace the acid rain issue:

1 The Thatcher Conservatives were ideologically committed to reductions in public spending on emission controls (Yearley 1992: 108).

2 The Conservative government contemplated privatising the nation's electricity industry and felt that costly anti-pollution requirements (Park 1987: 243) and a spate of future environmental regulations (Yearley 1992: 108) would make it less attractive to prospective buyers.

3 The Thatcher cabinet was embroiled in a continuing power struggle with the National Union of Mineworkers (NUM) during the early 1980s in which its prime policy arm was the CEGB. As a consequence, it was more concerned with stockpiling coal in case of a strike (which would lead to power cuts) than in worrying about the environmental effects of emissions from coal-burning plants (Boehmer-Christiansen and Shea 1991: 214–15).

3 One indicator of this is the exodus of researchers and research money to newer problems. For example, in Norway, a large greenhouse built on a remote hillside to demonstrate that streams lose their acidity when rainwater is intercepted and cleaned up was redeployed in the early 1990s to study the effects of greenhouse warming (Pearce 1990: 57).

8 BIODIVERSITY LOSS: THE SUCCESSFUL 'CAREER' OF A GLOBAL ENVIRONMENTAL PROBLEM

1   At the first official meeting to set research priorities in conservation biology in April 1988, many participants cited the need for aggressive conservation action rather than research as the highest priority (Tangley 1988: 444).

2   My chronology of these international conventions draws primarily on 'Annex 3: International legislation supporting conservation of biological diversity' in McNeely *et al.* (1990a).

3   As Mazur and Lee (1993: 701) note:

> The plights of these animals became salient through popular books, television documentaries such as those produced by the National Geographic Society, and news coverage, often of a few effective spokespersons including Jacques Cousteau, Brigitte Bardot, Roger Tory. Peterson and Jane Goodall, who usually specialized in a single type of animal.

4   The Northern spotted owl became one of 'the most celebrated and vili-fiedangered species' (Grumbine 1992: 144) in recent memory. With a habitat and geographic range which stretches the length of old growth forests from British Columbia to Northern California, protecting it under the Endangered Species Act implied a significant reduction in logging activities in the ancient forests. In the course of a decade of political and legal wrangling the Northern spotted owl became a symbol for some of the unrealistic features of the Act.

5   This appears to have been a two-way street. Not only did the fate of the dinosaurs provide a powerful magnet by which biodiversity activists could attract the attention of the public, but research on the immediate threat of extinction has proved useful in understanding what happened 245 million years ago. For example, Niles Eldredge, in writing his book *The Miner's Canary: Unraveling the Mysteries of Extinction* (1991), relied heavily on Edward Wilson's published data and arguments to examine the relationship between the mass extinctions of the geological past and the present-day biodiversity crisis (Eldredge 1992–3: 90).

6   Although formally separate, these international bodies maintain functionally close links. Boardman (1991: 109–10) claims that by the late 1970s the IUCN and the WWF had in effect been administratively fused, 'if not into a unified spearhead then at least into the twin prongs of a fondue fork'.

9 BIOTECHNOLOGY AS AN ENVIRONMENTAL PROBLEM: THE BOVINE GROWTH HORMONE CONFLICT

1   It should be noted, however, that the initial definition of genetic engineering as a potential hazard can be traced to the 1973 Gordon Conference on nucleic acids. A majority of the ninety scientists present at the conference voted to express their concerns over the newly emerging recombinant DNA research

in the form of letters to the National Academy of Sciences (NAS) and to *Science* (Krimsky 1979). The following year, a group of scientists representing the Committee on Recombinant DNA at the Academy sent their own letter to *Science* (26 July 1974). Known as the 'Berg letter', it outlined concerns over the dangers of genetically manipulated micro-organisms and the ability of existing research guidelines to ensure safety and called for a moratorium on genetic research (Plein 1990: 152). In Britain, the Berg letter was published on 19 July 1974, one week before its publication in the United States, in *Nature*. Rising concern over the issues raised in the letter led to the creation in 1976 of a UK central advisory committee under the title of the Genetic Manipulation Advisory Group (GMAG), an acronym which became an 'international byword' (Russell 1988: 113).

2   The genetically engineered ice nucleated (INA) bacteria developed by scientists at the University of California, Berkeley, and the biotechnology firm, Advanced Genetic Sciences, were said by opponents to threaten the environment in two ways: (1) most strains were known to be pathogenic (harmful) to a large list of plants including all the major agricultural crops in Northern California; (2) it was feared that field trials would significantly reduce the levels of naturally occurring INA bacteria in the upper atmosphere which centrally contribute to the formation of rain and snow. Between 1983 and 1986, the Foundation on Economic Trends (FET), a group headed by high-profile activist Jeremy Rifkin, filed a series of lawsuits to stop the release of ice minus bacteria. While these actions did not prevent the trials they were successful in slowing up the approval process and in generating national publicity (see Krimsky and Plough 1988).

3   Herbicide tolerance was a major issue at the Second European Network Meeting on Genetic Resources and Biotechnology held in Barcelona in June 1991. At this conference, seventy participants representing NGOs in fifteen countries agreed to coordinate a campaign against the research and production of herbicide resistant plants in Europe (Hindmarsh 1991: 203). Bora and Dobert (1992) present a case history of a German 'technology assessment' at which scientists, representatives from the herbicide industry, state officials and social movement (i.e. feminist, anti-nuclear, anti-capitalist) members attempted to find some common ground.

4   One scientist who actively supports the idea of 'burn-out' is David Kronfeld, Paul Mellon distinguished professor of agriculture and veterinary medicine at Virginia Tech. Kronfeld recently told an interviewer from *Feedstuffs*, an agri-business weekly, that informal reports from cooperating farmers have indicated that 'hypermetabolic syndrome' or 'BST burn-out' can show up at any time between forty and 600 days after bST administration was begun and that most poor milk responses are attributable to this burn-out (Carlson 1993).

5   The Green Revolution was a transformation in plant breeding that produced new high-yielding wheat and rice varieties. Touted as a solution to the chronic food shortages in the Third World, these new 'miracle grains' required the application of modern high-tech agricultural methods, notably heavy nitrogen fertiliser applications. While the new methods did increase food production, they also had a number of negative costs; notably, increased dependency on inputs from Western agro-corporations, rising individual debt levels among farmers, loss of genetic variability, bringing an increased susceptibility to crop

disease and insect infestation and pollution from intensive fertiliser application and pesticide use. For a good summary of the Green Revolution and its problems see Doyle (1985, Chapter 13).

6   It should be noted that Epstein's interest in this issue is not entirely of recent vintage. In a 1990 journal article, he concluded that regulatory agencies and producers have failed to demonstrate either the safety or efficacy of bST and that the use of milk hormones poses serious risks of adverse public health effects. The previous year he had made similar claims in a *Los Angeles Times* article (27 July 1989). See Krimsky 1991.

BIBLIOGRAPHY

Agyeman, J. (1990) 'A positive image', *Countryside Commission News* 45(3), September to October.

Albrecht, S. and Mauss, A. (1975) 'Environment', in A. Mauss *Social Problems as Social Movements*. Philadelphia: J.B. Lippincott Co.

Allen, R. (1992) *Waste Not, Want Not: The Production and Dumping of Toxic Waste*. London: Earthscan Publications Ltd.

Alpert, P. (1993) 'Support for biodiversity research from the U.S. Agency for International Development', *BioScience* 43(9): 628–31.

Altheide, D. (1976) *Creating Reality: How TV News Distorts Events*, Beverly Hills: Sage.

Alvo, R. (1987) 'Is the laughter of loons to be stilled on acid lakes?', *Canadian Geographic* 107(3): 46–50.

Anderson, A. (1993) 'Source–media relations: the production of the environmental agenda', in A. Hansen (ed.) *The Mass Media and Environmental Issues*. Leicester: Leicester University Press.

Andrews, E. L. (1991) 'Human threat ruled out in drug for cows', *New York Times*, 18 May: A-18.

Angier, N. (1994) 'Redefining diversity: biologists urge look beyond rain forests', *New York Times*, 29 November: B-5; B-9.

Aronoff, M. and Gunter, V. (1992) 'Defining disaster: local constructions for recovering in the aftermath of chemical contamination', *Social Problems* 9: 345–65.

Aronson, N. (1984) 'Science as a claims-making activity: implications for social problems research', in J. Schneider and J. I. Kitsuse (eds) *Studies in the Sociology of Social Problems*. Norwood, NJ: Ablex.

Ashworth, W. (1986) *The Late Great Lakes*. Toronto: Collins.

Avery, D. (forthcoming) *Saving the Planet with Pesticides and Plastic: The Environmental Triumph of High Yield Farming*, cited in *Feedstuffs* 65(48), 22 November 1993: 28.

Bagguley, P. (1992) 'Social change, the middle class and the emergence of "New Social Movements": a critical analysis', *The Sociological Review* 40(1): 26–48.

Bahro, R. (1984) *From Red to Green*. London: Verso.

Bailes, K. E. (1985) 'Critical issues in environmental history', in K. E. Bailes (ed.) *Environmental History: Critical Issues in Comparative Perspective*. Lantham, MD: University Press of America.

Bailey, R. (1993) *Ecoscam: The False Prophets of Ecological Apocalypse*. New York: St Martin's Press.

Barton, J. H. (1992) 'Biodiversity at Rio', *BioScience* 42(10): 773–6.

Baudrillard, J. (1983) *Simulations*. New York: Semiotext(e).

Bauman, Z. (1994) Review of M. Douglas, *Risk and Blame: Essays in Cultural Theory* [London: Routledge, 1992], *The British Journal of Sociology* 45(1): 143–4.

Baumann, E. A. (1989) 'Research rhetoric and the social construction of elder abuse', in J. Best (ed.) *Images of Issues: Typifying Contemporary Social Problems*. New York: Aldine de Gruyter.

Beck, U. (1992) *The Risk Society*. London: Sage.

Benedick, R. E. (1991) *Ozone Diplomacy: New Directions in Safeguarding the Planet*. Cambridge and London: Harvard University Press.

Benford, R. D. (1993) 'Frame disputes within the nuclear disarmament movement', *Social Forces* 71(3): 677–701.

Benton, T. (1991) 'Biology and social science: why the return of the repressed should be given a (cautious) welcome', *Sociology* 25(1): 1–29.

Benton, T. (1994) 'Biology and social theory in the environmental debate', in M. Redclift and T. Benton (eds) *Social Theory and the Global Environment*. London and New York: Routledge.

Benton, T. and Redclift, M. (1994) 'Introduction', in M. Redclift and T. Benton (eds) *Social Theory and the Global Environment*. London and New York: Routledge.

Berger, P. (1986) *The Capitalist Revolution*. New York: Basic Books.

Best, J. (1987) 'Rhetoric in claims-making', *Social Problems* 34(2): 101–21.

Best, J. (ed.) (1989a) *Images of Issues: Typifying Contemporary Social Problems*. New York: Aldine de Gruyter.

Best, J. (1989b) 'Afterword: extending the constructionist perspective: a conclusion – and an introduction', in Best (1989a), *Images of Issues: Typifying Contemporary Social Problems*. New York: Aldine de Gruyter.

Best, J. (1993) 'But seriously folks: the limitations of the strict constructionist interpretation of social problems', in J. A. Holstein and G. Miller (eds) *Reconsidering Social Constructionism: Debates in Social Problems Theory*. New York: Aldine de Gruyter.

Bierstedt, R. (1981) *American Sociological Theory: A Critical History*. New York: Academic Press.

Bird, E. A. (1987) 'The social construction of nature: theoretical approaches to the history of environmental problems', *Environmental Review* 11: 255–64.

Blakeslee, A. M. (1994) 'The rhetorical construction of novelty: presenting claims in a letters forum', *Science Technology and Human Values* 19(1): 88–100.

Blowers, A. (1993) 'Environmental policy: the quest for sustainable development', *Urban Studies* 30(4/5): 775–96.

Blowers, A., Lowry, D. and Solomon, B. D. (1991) *The International Politics of Nuclear Waste*. London: Macmillan.

Boardman, R. (1991) *International Organization and the Conservation of Nature*. London: The Macmillan Press Ltd.

Bocking, S. (1993) 'Conserving nature and building a science: British ecologists and the origins of the Nature Conservancy', in M. Shortland (ed.) *Science and Nature: Essays in the History of the Environmental Sciences*. Oxford: British Society for the History of Science.

Boehmer-Christiansen, S. and Skea, J. (1991) *Acid Politics: Environmental and Energy Politics in Britain and Germany*. London and New York: Belhaven Press.

Boorstin, D. (1964) *The Image: A Guide to Pseudo-Events in America*. New York: Harper & Row.

Bora, A. and Dobert, R. (1992) 'Prerequisites of procedural rationality: notes on a German technology assessment of transgenetic plants with herbicide resistance', Paper presented to the Symposium, 'Current Developments in Environmental Sociology', International Sociological Association, Woudschoten, The Netherlands.

Borgatta, E. F. (1994) 'Sociology and the reality of the press on environmental resources', in W. V. D'Antonio, M. Sasaki and Y. Yonebayashi (eds) *Ecology, Society and the Quality of Social Life*. New Brunswick, NJ and London: Transaction Publishers.

Borie, L. (1987) 'Are sugar maples declining?', *American Forests*, November/December: 26–8; 66–70.

Bosso, C. J. (1987) *Pesticides and Politics: The Life Cycle of a Public Issue*. Pittsburgh, PA: University of Pittsburgh Press.

Boyle, R. H. and Boyle, R. A. (1983) *Acid Rain*. New York: Schocken.

Bramble, B. J. and Parker, G. (1992) 'Non-governmental organizations and the making of U.S. international environmental policy', in A. Hurrell and B. Kingsbury (eds) *The International Politics of the Environment*. Oxford: Clarendon Press.

Bramwell, A. (1989) *Ecology in the 20th Century: A History*. New Haven and London: Yale University Press.

Brechin, S. R. and Kempton, W. (1994) 'Global environmentalism: a challenge to the postmaterialism thesis?', *Social Science Quarterly* 75(2): 245–69.

Breyman, S. (1993) 'Knowledge as power: ecology movements and global environmental problems', in R. D. Lipschutz and K. Conca (eds) *The State and Social Power in Global Environmental Politics*. New York: Columbia University Press.

Brookes, S. H., Jordan, A. G., Kimber, R. H. and Richardson, J. J. (1976) 'The growth of the environment as a political issue in Britain', *British Journal of Political Science* 6: 245–55.

Brown, L. R. (1986) 'And today we're going to talk about biodiversity ... that's right, biodiversity', in E. O. Wilson (ed.) *Biodiversity*. Washington, DC: National Academy Press.

Browne, W. P. (1987) 'Bovine growth hormone and the politics of uncertainty: fear and loathing in a transitional agriculture', *Agriculture and Human Values* 4(1): 75–80.

Browne, W. P. and Hamm, L. G. (1990) 'Political choices, social values, and the economics of biotechnology: a lesson from the dairy industry', in D. J. Webber (ed.) *Biotechnology: Assessing Social Impacts and Policy Implications*. Westport, CT: Greenwood Press.

Bryant. B. and Mohai, P. (1992) 'Introduction', in B. Bryant and P. Mohai (eds) *Race and the Incidence of Environmental Hazards: A Time for Discourse*. Boulder, CO: Westview Press.

Bullard, R. D. (1990) *Dumping in Dixie: Race, Class and Environmental Quality*. Boulder, CO: Westview Press.

Burgess, J. and Harrison, C. M. (1993) 'The circulation of claims in the cultural politics of environmental change', in A. Hansen (ed.) *The Mass Media and Environmental Issues*. Leicester: Leicester University Press.

Buttel, F.H. (1986) 'Sociology and the environment: the winding road toward human ecology', *International Social Science Journal* 38(3): 337–56.

Buttel, F.H. (1987) 'New directions in environmental sociology', *Annual Review of Sociology* 13: 465–88.

Buttel, F. and Taylor, P. (1992) 'Environmental sociology and global environmental change: a critical assesment', *Society and Natural Resources* 5: 211–30.

Buttel, F. and Taylor, P. (1994) 'Environmental sociology and global environmental change: a critical assessment', in M. Redclift and T. Benton (eds) *Social Theory and the Global Environment*. London and New York: Routledge.

Buttel, F.H. and Humphrey, C. (forthcoming) 'Sociological theory and the natural environment', in R. E. Dunlap and W. Michelson (eds) *Handbook of Environmental Sociology*. Westport, CT: Greenwood Press.

Cable, S. and Cable, C. (1995) *Environmental Problems Grassroots Solution: The Politics of Grassroots Environmental Conflict*. New York: St Martin's Press.

Calhoun, C. and Ritzer, G. (1993) *Social Problems*. New York: McGraw-Hill.

Canada [House of Commons] (1981) Still Waters: Report by the Subcommittee on Acid Rain, Executive Summary. Ottawa: Ministry of Supply and Services.

Cantrill, J. G. (1992) 'Understanding environmental advocacy: interdisciplinary research and the role of cognition', *The Journal of Environmental Education* 24: 35–42.

Capek, S. M. (1993) 'The "environmental justice" frame: a conceptual discussion and an application', *Social Problems* 40: 5–24.

Capuzza, J. (1992) 'A critical analysis of image management within the environmental movement', *The Journal of Environmental Education* 24: 9–14.

Carlson, G. S. (1993) 'Debate over dairy cow "burnout" restarted after previous column', *Feedstuffs*, 29 November: 2.

Carlson, G. S. (1994) 'Link suggested between BST, breast cancer in women', *Feedstuffs*, 14 March: 12.

Carson, R. (1962) *Silent Spring*. Boston: Houghton Mifflin.

Cataldo, E. (1992) 'Acid rain policy in the United States: an explanation of Canadian influence', *The Social Science Journal* 29(4): 395–409.

Catton, W. R. Jr. (1994) 'Foundations of human ecology', *Sociological Perspectives* 37(1): 75–95.

Catton, W. R. Jr. and Dunlap, R. E. (1989) 'Competing functions of the environment: living space, supply depot and waste repository', Paper presented at the annual meeting of the Rural Sociological Society, Salt Lake City, Utah.

Chavez, C. (1993) 'Farm workers at risk', in R. Hofrichter (ed.) *Toxic Struggles: The Theory and the Practice of Environmental Justice*. Philadelphia, PA: New Society Publishers.

Chiras, D. D. (1988) *Environmental Science: A Framework for Decision Making*, Menlo Park, CA: The Benjamin Cummings Publishing Co Inc.

Churchill, W. and La Duke, W. (1985) 'Radioactive colonization and the Native American', *Socialist Review* 15: 95–120.

Cittadino, E. (1993) 'The failed promise of human ecology', in M. Shortland (ed.) *Science and Nature: Essays in the History of the Environmental Sciences*. Oxford: British Society for the History of Science.

Clarke, D. (1981) 'Second-hand news: production and reproduction at a major

Ontario television station', in L. Salter (ed.) *Communication Studies in Canada.* Toronto: Butterworth.

Clarke, D. (1992) 'Constraints of television news production: the example of story geography', in M. Grenier (ed.) *Critical Studies of Canadian Mass Media.* Toronto: Butterworth.

Clarke, L. (1988) 'Explaining choices among technological risks', *Social Problems* 35(1): 22–35.

Clarke, L. and Short, J. F. Jr. (1993) 'Social organization and risk: some current controversies', *Annual Review of Sociology* 19: 375–99.

Clements, F. E. (1905) *Research Methods in Ecology.* Lincoln: University Publishing Company. Reprinted by Arno Press (1977).

Cline, W. R. (1992) *The Economics of Global Warming.* Washington, DC: Institute for International Economics.

Coleman, J. W. and Cressey, D. R. (1980) *Social Problems.* New York: Harper & Row.

Collingridge, D. and Reeve, C. (1986) *Science Speaks to Power: The Role of Experts in Policy Making.* London: Frances Pinter.

Commoner, B. (1963) *Science and Survival.* New York: Viking.

Commoner, B. (1971) *The Closing Circle.* New York: Bantam Books.

Corbett, J. B. (1993) 'Atmospheric ozone: a global or local issue? Coverage in Canadian and U.S. newspapers', *Canadian Journal of Communication* 18: 81–7.

Cormier, J. (1994) 'Doomsday, in handy disc form', *Equinox* 78, November/December: 4.

Corner, J. and Richardson, K. (1993) 'Environmental communication and the contingency of meaning: a research note', in A. Hansen (ed.) *The Mass Media and Environmental Issues.* Leicester: Leicester University Press.

Cotgrove, S. F. (1982) *Catastrophe or Cornucopia: The Environment, Politics and the Future.* Chicester, Sussex: John Wiley.

Cotgrove, S. (1991) 'Sociology and the environment: Cotgrove replies to Newby', *Network* (British Sociological Association) 51, October.

Cotgrove, S. F. and Duff, A. (1981) 'Environmentalism, values and social change', *British Journal of Sociology* 32(1): 92–110.

Cottle, S. (1993) 'Mediating the environment: modalities of TV news', in A. Hansen (ed.) *The Mass Media and Environmental Issues.* Leicester: Leicester University Press.

Covello, V. T. and Johnson, B. B. (1987) 'The social and cultural construction of risk: issues, methods and case studies', in B. B. Johnson and V. T. Covello (eds) *The Social and Cultural Construction of Risk: Essays on Risk Selection and Perception.* Dordrecht, Holland: D. Reidel Publishing Company.

Cowling, E. B. (1982) 'Acid precipitation in historical perspective', *Environmental Science and Technology* 16: 110A–23A.

Cracknell, J. (1993) 'Issue arenas, pressure groups and environmental agendas', in A. Hansen (ed.) *The Mass Media and Environmental Issues.* Leicester: Leicester University Press.

Crenson, M. A. (1972) *The Un-politics of Air Pollution.* Baltimore, VA: Johns Hopkins University Press.

Cronor, C. F. (1993) *Regulating Toxic Substances: A Philosophy of Science and Law.* New York: Oxford University Press.

Cylke, F. K. Jr. (1993) *The Environment*. New York: HarperCollins College Publishers.

Dake, K. (1992) 'Myths of nature: culture and the social construction of risk', *Journal of Social Issues* 48(4): 21–37.

Daley, P. and O'Neill, D. (1991) 'Sad is too mild a word: press coverage of the *Exxon Valdez* oil spill', *Journal of Communication* 41: 42–57.

Daniel, P. (1986) *Breaking the Land: The Transformation of Cotton, Tobacco and Rice Cultures Since 1880*. Champaign, IL: University of Illinois Press.

Dickens, P. (1992) *Society and Nature: Towards a Greener Social Theory*. Hemel Hempstead: Harvester Wheatsheaf.

Dietz, T. R. and Rycroft, R. W. (1987) *The Risk Professionals*. New York: Russell Sage Foundation.

Dietz, T.R., Frey, R. S. and Rosa, E. (forthcoming) 'Risk, technology and society', in R. E. Dunlap and W. Michelson (eds) *Handbook of Environmental Sociology*. Westport, CT: Greenwood Press.

Dorfman, A. (1992) 'Sideshows galore', *Time* 139(22), 1 June: 43.

Dotto, L. (1993) 'Global warming: dealing with uncertainty', *The Globe & Mail*, 13 November: D-5.

Doughty, R. W. (1975) *Feather Fashions and Bird Preservation: A Study in Nature Protection*. Berkeley and Los Angeles, CA: University of California Press.

Douglas, M. A. and Wildavsky, A. (1982) *Risk and Culture: An Essay on the Selection of Technological and Environmental Dangers*. Berkeley, CA: University of California Press.

Downs, A. (1972) 'Up and down with ecology – the "issue–attention" cycle', *The Public Interest* 28: 38–50.

Doyle, J. (1985) *Altered Harvest: Agriculture, Genetics and the Fate of the World's Food Supply*. New York: Penguin Books.

Duncan, O. D. (1961) 'From social system to ecosystem', *Sociological Inquiry* 31: 140–9.

Dunk, T. (1994) 'Talking about trees: environment and society in forest workers' culture', *Canadian Review of Sociology and Anthropology* 31: 14–34.

Dunlap, R. E. (1993) 'From environmental problems to ecological problems', in C. Calhoun and George Ritzer (eds) *Social Problems*. New York: McGraw-Hill.

Dunlap, R. E. (1994) 'Limitations of the social constructionist approach to environmental problems', Paper presented to the XIIIth World Congress of Sociology, Bielefeld, Germany.

Dunlap, R. E. and Catton, W. R. Jr. (1979) 'Environmental sociology', *Annual Review of Sociology* 5: 243–73.

Dunlap, R. E. and Catton, W. R. Jr. (1983) 'What environmental sociologists have in common', *Sociological Inquiry* 33: 113–35.

Dunlap, R. E. and Catton, W. R. Jr. (1992/3) 'Towards an ecological sociology: the development, current status and probable future of environmental sociology', *The Annals of the International Institute of Sociology* 3 (New Series): 263–84.

Dunlap, R. E. and Scarce, R. (1990) 'The polls: poll trends, environmental problems and protection', *Public Opinion Quarterly* 55: 651–6.

Dunlap, T. R. (1991) 'Organization and wildlife preservation: the case of the whooping crane in North America', *Social Studies of Science* 21: 197–221.

Dunwoody, S. and Griffin, R. L. (1993) 'Journalistic strategies for reporting long-term environmental issues: a case study of three Superfund sites', in

A. Hansen (ed.) *The Mass Media and Environmental Issues.* Leicester: Leicester University Press.

Edelman, M. (1964) *The Symbolic Use of Politics.* Urbana: University of Illinois Press.

Edelman, M. (1977) *Political Language: Words That Succeed and Policies That Fail.* New York: Academic Press.

Ehrlich, P.R. (1968) *The Population Bomb.* New York: Ballantine Books.

Ehrlich, P. R. and Ehrlich, A. (1981) *Extinction: The Causes and Consequences of the Disappearance of Species.* New York: Random House.

Einsiedal, E. and Coughlan, E. (1993) 'The Canadian press and the environment: reconstructing a social reality', in A. Hansen (ed.) *The Mass Media and Environmental Issues.* Leicester: Leicester University Press.

Eisner, T. (1989–90) 'Prospecting for nature's chemical riches', *Issues in Science and Technology* 6: 31–4.

Eisner, T. and Beiring, E. A. (1994) 'Biotic Exploration Fund – protecting biodiversity through chemical prospecting', *BioScience* 44(2): 95–8.

Eldredge, N. (1991) *The Miner's Canary: Unraveling the Mysteries of Extinction.* Englewood Cliffs, NJ: Prentice-Hall.

Eldredge, N. (1992–3) 'Confronting the skeptics' (Review of E. O. Wilson, *The Diversity of Life* [Cambridge, MA: Harvard University Press, 1992]), *Issues in Science and Technology* 9(2): 90–2.

Ellis, D. (1989) *Environments at Risk: Case Histories of Impact Assessment.* Berlin/Heidelberg/New York: Springer-Verlag.

Enloe, C. H. (1975) *The Politics of Pollution in a Comparative Perspective: Ecology and Power in Four Nations.* New York: David McKay Co Inc.

enviro (1993) 'Acidification still a major cause for concern', *enviro: International Magazine on the Environment* 16, December: 22–3.

Environmental Conservation (1993) 'Editor's note: the biodiversity treaty', *Environmental Conservation* 20(3): 277.

Enzenberger, H. M. (1979) 'A critique of political ecology', in A. Cockburn and J. Ridgeway (eds) *Political Ecology.* New York: Quadrangle.

Epstein, B. (1993) 'Ecofeminism and grass-roots environmentalism in the United States', in R. Hofrichter (ed.) *Toxic Struggles: The Theory and Practice of Environmental Justice.* Philadelphia, PA: New Society Publishers.

Epstein, S. (1989) 'Growth hormone would endanger milk', *Los Angeles Times*, 27 July.

Epstein, S. (1990) 'Potential public health hazards of biosynthetic milk hormones', *International Journal of Health Services* 24(1): 73–84.

Evans, D. (1992) *A History of Nature Conservation in Britain.* London and New York: Routledge.

Eyerman, R. and Jamison, A. (1989) 'Environmental knowledge as an organizational weapon: the case of Greenpeace', *Social Science Information* 28(1): 99–119.

Faber, D. and O'Connor, J. (1993) 'Capitalism and the crisis of environmentalism', in R. Hofrichter (ed.) *Toxic Struggles: The Theory and Practice of Environmental Justice.* Philadelphia, PA: New Society Publishers.

Featherstone, M. (1988) 'In pursuit of the postmodern: an introduction', *Theory, Culture & Society* 5(2–3): 195–215.

Feedstuffs (1993) 'Montsanto promotion kit highlights safety, economics', *Feedstuffs*, 15 November: 4.

Feedstuffs (1994a) 'Europe slips far behind BST adopters', *Feedstuffs*, 7 March: 8.

Feedstuffs (1994b) 'BST labeling restrictions under consideration nationwide', *Feedstuffs*, 9 May: 11.

Feedstuffs (1994c) 'Truth is anecdote for wild BST rhetoric', *Feedstuffs*, 17 October: 10.

Firey, W. (1947) *Land Use in Central Boston*. Cambridge, MA: Harvard University Press.

Fishman, M. (1980) *Manufacturing the News*. Austin: University of Texas Press.

Fletcher, F. J. and Stahlbrand, L. (1992) 'Mirror or participant? The news media and environmental policy', in R. Boardman (ed.) *Canadian Environmental Policy: Ecosystems, Politics and Process*. Toronto: Oxford University Press.

Fowler, C., Lachkovics, E., Mooney, P. and Shand, H. (1988) 'The laws of life: another development and the new biotechnologies', *Development Dialogue* 1–2: 1–350.

Fox, M. W. (1986) 'Effects of growth hormone on cows', *Science* 223, 19 September: 1247.

Fox, S. R. (1981) *John Muir and His Legacy: The American Conservation Movement*. Boston, MA: Little, Brown & Co.

Franck, I. and Brownstone, D. (1992) *The Green Encyclopedia*. New York: Prentice-Hall General Reference.

Freudenburg, W. R. and Gramling, R. (1989) 'The emergence of environmental sociology: contributions of Riley, E. Dunlop and William R. Catton, Jr.', *Sociological Inquiry* 59(4): 439–52.

Freudenburg, W. R. and Pastor, S. (1992) 'Public responses to technological risks: toward a sociological perspective', *The Sociological Quarterly* 33: 389–412.

Friedman, S. M. (1983) 'Environmental reporting: a problem child of the media', *Environment* 25(10): 24–9.

Friedman, S. M. (1984) 'Environmental reporting: before and after TMI', *Environment* 26(10): 4–5; 34.

Fumento, M. (1993) *Science Under Siege: Balancing Technology and the Environment*. New York: William Morrow.

Gamson, W.A. and Modigliani, A. (1989) 'Media discourse and public opinion on nuclear power', *American Journal of Sociology* 95: 1–37.

Gamson, W. A., Croteau, D., Haynes, W. and Sasson, T. (1992) 'Media images and social construction of reality', *Annual Review of Sociology* 18: 373–93.

Gamson, W. A. and Wolfsfeld, G. (1993) 'Movements and media as interacting systems', *Annals* (AAPSS) 528: 114–25.

Gans, H. J. (1979) *Deciding What's News: A Study of CBS Evening News, NBC Nightly News, Newsweek and Time*. New York: Pantheon Books.

Geddes, J. (1994) 'Copps faces first major challenge: setting environmental regulations tough task', *The Financial Post* (Toronto), 18 June: 8.

Giddens, A. (1990) *The Consequences of Modernity*. Cambridge: Polity Press.

Giddens, A. (1991) *Modernity and Self-Identity*. Cambridge: Polity Press.

Gitlin, T. (1980) *The Whole World Is Watching: Mass Media in the Making or Unmaking of the New Left*. Berkeley and Los Angeles, CA: University of California Press.

Gooderham, M. (1993) 'Battle lines drawn over bovine hormone: coalition to launch campaign against food biotechnology by picketing fast-food chain', *The Globe & Mail*, 17 April: A-4.

Gooderham, M. (1994a) 'Scientific group urges species inventory', *The Globe & Mail*, 23 February: A-12.

Gooderham, M. (1994b) 'Debate over milk hormone heats up', *The Globe & Mail*, 14 February: A-5.

Gorham, E. (1955) 'On the acidity and salinity of rain', *Geochimica et Cosmochimica Acta* 7: 231–9.

Gorrie, P. (1986) 'Acid rain fighter: the story of one man's persistent efforts to get us to stop polluting the environment', *Canadian Geographic* 106(5): 8–17.

Gottlieb, R. (1993) *Forcing the Spring: The Transformation of the American Environmental Movement*. Washington, DC: Island Press.

Gould, K. A., Weinberg, A. S. and Schnaiberg, A. (1993) 'Legitimating impatience: Pyrrhic victories of the modern environmental movement', *Qualitative Sociology* 16(3): 207–46.

Gould, R. (1985) *Going Sour: Science and Politics of Acid Rain*. Boston, MA: Birkhauser.

Greenpeace (1991) *Toxic Threat to Indian Lands*. Washington, DC: Greenpeace.

Greider, T. and Garkovitch, L. (1994) 'Landscapes: the social construction of nature and the environment', *Rural Sociology* 59(1): 1–24.

Griffin, S. (1978) *Women and Nature: The Roaring Inside Her*. New York: Harper & Row.

Grove-White, R. (1993) 'Environmentalism: a new moral discourse for technological society?', in K. Milton (ed.) *Environmentalism: The View From Anthropology*. London and New York: Routledge.

Grumbine, R. E. (1992) *Ghost Bears: Exploring the Biodiversity Crisis*. Washington, DC: Island Press.

Gusfield, J. R. (1981) *The Culture of Public Problems: Drinking Driving and the Symbolic Order*. Chicago: University of Chicago Press.

Gusfield, J. R. (1984) 'On the side: practical action and social constructivism in social problems theory', in J. W. Schneider and J. I. Kitsuse (eds) *Studies in the Sociology of Social Problems*. Norwood, NJ: Ablex.

Haas, P. (1990) *Saving the Mediterranean: The Politics of International Environmental Cooperation*. New York: Columbia University Press.

Haas, P. (1992) 'Obtaining international protection through epistemic consensus', in I. H. Rowlands and M. Greene (eds) *Global Environmental Change and International Relations*. Basingstoke: Macmillan.

Habermas, J. (1987) *The Theory of Communicative Action*, Vol. 2. Cambridge: Polity Press.

Hagen, J. B. (1992) *An Entangled Bank: The Origins of Ecosystem Ecology*. New Brunswick, NJ: Rutgers University Press.

Hager, C. (1993) 'Citizen movements and technological policymaking in Germany', *The Annals* (AAPSS) 528: 42–55.

Halfmann, J. and Japp, H. P. (1993) 'Modern social movements as active risk observers: a systems–theoretical approach to collective action', *Social Science Information* 32(3): 427–46.

Hannigan, J. A. (1985) *Laboured Relations: Reporting Industrial Relations News in Canada*. Toronto: Centre for Industrial Relations, University of Toronto.

Hannigan, J. A. (1995) 'Sociology and the environment', in R. Brym (ed.) *New Society: Sociology for the 21st Century*. Toronto: Harcourt Brace Canada.

Hansen, A. (1991) 'The media and the social construction of the environment', *Media, Culture & Society* 13(4): 443–58.

Hansen, A. (1993a) 'Introduction', in A. Hansen (ed.) *The Mass Media and Environmental Issues*. Leicester: Leicester University Press.

Hansen, A. (1993b) 'Greenpeace and press coverage of environmental issues', in A. Hansen (ed.) *The Mass Media and Environmental Issues*. Leicester: Leicester University Press.

Hardin, G. (1978) 'The tragedy of the commons', *Science* 162: 1241–52.

Harries Jones, P. (1993) 'Between science and shamanism: the advocacy of environmentalism in Toronto', in K. Milton (ed.) *Environmentalism: The View from Anthropology*. London and New York: Routledge.

Harrison, K. and Hoberg, G. (1991) 'Setting the environmental agendas in Canada and the United States: the cases of dioxin and radon', *Canadian Journal of Political Science* 24: 3–27.

Harrison, K. and Hoberg, G. (1994) *Risk, Science, and Politics: Regulating Toxic Substances in Canada and the United States*. Montreal and Kingston: McGill–Queen's University Press.

Hart, D. M. and Victor, D. G. (1993) 'Scientific elites and the making of U.S. policy for climate change research, 1957–1974', *Social Studies of Science* 23: 643–80.

Harvey, D. (1974) 'Population, resources, and the ideology of science', *Economic Geography* 50: 256–77.

Hawkins, A. (1993) 'Contested ground: international environmentalism and global climate change', in R. D. Lipschutz and K. Conca (eds) *The State and Social Power in Global Environmental Politics*. New York: Columbia University Press.

Hays, S. (1959) *Conservation and the Gospel of Efficiency: The Progressive Conservation Movement*. Cambridge, MA: Harvard University Press.

Heimer, C. (1988) 'Social structure, psychology and the estimation of risk', *Annual Review of Sociology* 14: 491–519.

Higgins, R. R. (1993) 'Race and environmental equity: an overview of the environmental justice issue in the policy process', *Polity* 26(2): 281–300.

Higham, J. (1963) *Strangers in the Land: Patterns of American Nativism, 1860–1925*. New York: Atheneum.

Hilgartner, S. (1992) 'The social construction of risk objects: or how to pry open networks of risk', in J. F. Short Jr. and L. Clarke (eds) *Organizations, Uncertainties and Risk*. Boulder, CO: Westview Press.

Hilgartner, S. and Bosk, C. L. (1988) 'The rise and fall of social problems: a public arenas model', *American Journal of Sociology* 94(1): 53–78.

Hindmarsh, R. (1991) 'The flawed "sustainable" promise of genetic engineering', *The Ecologist* 21(5): 196–205.

Hirsch, J. and Roth, R. (1986) *Das Neue Geisicht des Kapitalismus: Vom Fordismus zum Post-Fordismus* (The New Face of Capitalism: From Fordism to Post-Fordism). Hamburg: VSA-Verlag.

Hochberg, L. (1980) 'Environmental reporting in boomtown Houston', *Columbia Journalism Review* 19: 71–4.

Hofrichter, R. (1993) 'Introduction', in R. Hofrichter (ed.) *Toxic Struggles: The Theory and Practice of Environmental Justice*. Philadelphia, PA: New Society Publishers.

Hofstadter, R. (1959) *Social Darwinism in American Thought*. New York: George Braziller.

Holstein, J. A. and Miller, G. (eds) (1993) *Reconsidering Social Constructionism: Debates in Social Problems Theory.* New York: Aldine de Gruyter.

Hornsby, M. (1990) 'Milk-yield hormone for cows faces ban', *The Times* (London), 15 September: 4.

Howard, L. E. (1953) *Sir Albert Howard in India.* London: Faber & Faber.

Howenstine, E. (1987) 'Environmental reporting: shift from 1970 to 1982', *Journalism Quarterly* 64(4): 842–6.

Huber, J. (1982) *Die Verlorene Unschuld der Okologie: Neue Technologien und Super-industrielle Entwicklung* (The Lost Innocence of Ecology: New Technologies and Superindustrial Development). Frankfurt am Main: Fischer.

Huber, J. (1985) *Die Regenbogengesellschaft: Okologie und Sozialpolitik* (The Rainbow Society: Ecology and Social Policy). Frankfurt am Main: Fischer.

Hunt, S. A., Benford, R. D. and Snow, D. A. (1994) 'Identity fields: framing processes and the social construction of movement identities', in E. Larana, H. Johnston and J. R. Gusfield (eds) *New Social Movements: From Ideology to Identity.* Philadelphia: Temple University Press.

Ibarra, P. R. and Kitsuse, J. I. (1993) 'Vernacular constituents of moral discourse: an interactionist proposal for the study of social problems', in J. A. Holstein and G. Miller (eds) *Reconsidering Social Constructionism: Debates in Social Problems Theory.* New York: Aldine de Gruyter.

Inglehart, R. (1971) 'The silent revolution in Europe: intergenerational change in post-industrial societies', *American Political Science Review* 65: 991–1017.

Inglehart, R. (1977) *The Silent Revolution: Changing Values and Political Styles Among Western Publics.* Princeton, NJ: Princeton University Press.

Inglehart, R. (1990) *Culture Shift in Advanced Industrial Society.* Princeton, NJ: Princeton University Press.

Inkeles, A. and Smith, D. H. (1974) *Becoming Modern: Individual Change in Six Developing Countries.* Cambridge, MA: Harvard University Press.

Isaacs, C. (1994) 'Biodiversity may spur profits', *The Financial Post* (Toronto), 28 January: 17.

Jameson, F. (1984) 'Postmodernism or the cultural logic of late capitalism', *New Left Review* 146: 53–93.

Jasanoff, S. (1986) *Risk Management and Political Culture: A Comparative Study of Science in the Policy Context.* New York: Russell Sage Foundation.

Jasanoff, S. (1990) *The Fifth Branch: Science Advisers as Policymakers.* Cambridge, MA and London: Harvard University Press.

Jehlicka, P. (1992) 'Environmentalism in Europe?' Paper presented to the British Sociological Association Conference, April.

Jenness, V. (1993) *The Prostitutes' Rights Movement in Perspective.* Hawthorne, NY: Aldine de Gruyter.

Jonassen, C. T. (1949) 'Cultural variables in the ecology of an ethnic group', *American Sociological Review* 14: 32–41.

Kalter, R. R. (1985) 'The new biotech agriculture: unforeseen economic conse-quences', *Issues in Science and Technology* 2: 122–33.

Kaminstein, D. S. (1988) 'Toxic talk', *Social Policy* 19(2): 5–10.

Katz, E. and Lazarsfeld, P. (1955) *Personal Influence: The Part Played by People in the Flow of Mass Communication.* Glencoe, IL: Free Press.

Kawar, A. (1989) 'Issue definition, democratic participation, and genetic engineer-ing', *Policy Studies Journal* 17(4): 733–43.

Kellert, S. R. (1986) 'Social and perceptual factors in the preservation of animal species', in B. G. Norton (ed.) *The Preservation of Species: The Value of Biological Diversity*. Princeton, NJ: Princeton University Press.

Kellner, D. (1988) 'Postmodernism as social theory: some challenges and problems', *Theory, Culture & Society* 5(2–3): 239–69.

Kenney, M. (1986) *Biotechnology: The University–Industrial Complex*. New Haven: Yale University Press.

Killingsworth, M. J. and Palmer, J. S. (1992) *Ecospeak: Rhetoric and Environmental Politics in America*. Carbondale and Edwardsville, IL: Southern Illinois University Press.

Kingdon, J. W. (1984) *Agendas, Alternatives and Public Policies*. Boston, MA: Little, Brown & Co.

Kitsuse, J. I. and Spector, M. (1981) 'The labeling of social problems', in E. Rubington and M. S. Weinberg (eds) *The Study of Social Problems*. New York: Oxford University Press.

Kitsuse, J. I., Murase, A. E. and Yamamura, Y. (1984) 'The emergence and institutionalization of an educational problem in Japan', in J. W. Schneider and J. I. Kitsuse (eds) *Studies in the Sociology of Social Problems*. Norwood, NJ: Ablex.

Kleinman, D. L. and Kloppenburg, J. Jr. (1991) 'Aiming for the discursive high ground', *Sociological Forum* 6(3): 427–47.

Knorr-Cetina, K. D. (1983) 'The ethnographic study of scientific work: towards a constructivist interpretation of science', in K. Knorr-Cetina and M. Mulkay (eds) *Science Observed: Perspectives on the Social Study of Science*. London: Sage.

Kowalok, M. (1993) 'Research lessons from acid rain, ozone depletion and global warming', *Environment* 35(6): 13–20; 35–8.

Kriesi, H. (1989) 'New social movements and the new class in the Netherlands', *American Journal of Sociology* 94(5): 1078–116.

Krimsky, S. (1979) 'Regulating recombinant DNA research', in D. Nelkin (ed.) *Controversy: Politics of Technical Decisions*. Beverly Hills, CA: Sage.

Krimsky, S. (1991) *Biotechnics and Society: The Rise of Industrial Genetics*. New York: Praeger.

Krimsky, S. and Plough, A. (1988) *Environmental Hazards: Communicating Risks as a Social Process*. Dover, MA: Auburn House Publishing Company.

Kuchler, F., McClelland, J. and Offutt, S. E. (1990) 'Regulatory experience with food safety: social choice implications for recombinant DNA-derived animal growth hormones', in D. J. Weber (ed.) *Biotechnology: Assessing Social Impacts and Policy Implications*. Westport, CT: Greenwood Press.

Kunst, M. and Witlox, N. (1993) 'Communication and the environment', *Communication Research Trends* 13: 1–31.

Kwa, C. (1993) 'Radiation ecology, systems ecology and the management of the environment', in M. Shortland (ed.) *Science and Nature: Essays in the History of the Environmental Sciences*. Oxford: British Society for the History of Science.

Lacey, C. and Longman, D. (1993) 'The press and public access to the environment and development debate', *The Sociological Review* 41(2): 207–43.

Lash, S. and Urry, J. (1994) *Economies of Signs and Space*. London: Sage.

Lash, S. and Wynne, B. (1992) 'Introduction' to U. Beck, *The Risk Society*. London: Sage.

Laska, S. B. (1993) 'Environmental sociology and the state of discipline', *Social Forces* 72(1): 1–17.

Latour, B. and Woolgar, S. (1986) *Laboratory Life: The [Social] Construction of Scientific Facts*, revised edn. Princeton, NJ: Princeton University Press.

Laws, J. L. and Schwartz, P. (1977) *Sexual Scripts: The Social Construction of Female Sexuality*, Hinsdale, IL: The Dryden Press.

Lawson, S. (1994) 'Farm widow refights old pipeline foe', *The Toronto Star*, 22 January: C-6.

Lear, L. K. (1993) 'Rachel Carson's *Silent Spring*', *Environmental History Review* 17: 23–48.

Lee, C. (ed.) (1992) *Proceedings, First National People of Color Environmental Leadership Summit*. New York: United Church of Christ Commission for Racial Justice, December.

Lee, D. C. (1980) 'On the Marxian view of the relationship between man and nature', *Environmental Ethics* 2: 3–16.

Leopold, A. (1949) *A Sand County Almanac*. New York: Oxford University Press.

Lerner, D. (1958) *The Passing of Traditional Society: Modernizing the Middle East*. Glencoe, IL: The Free Press.

Lewis, M. W. (1992) *Green Delusions: An Environmental Critique of Radical Environmentalism*. Durham, NC: Duke University Press.

Liazos, A. (1989) *Sociology: A Liberating Perspective*, 2nd edn. Boston: Allyn & Bacon.

Liberatore, A. (1992) 'Facing global warming: the interactions between science and policy making in the European community', Paper presented at the Symposium, 'Current Developments in Environmental Sociology', International Sociological Association Thematic Group on 'Environment and Society', Woudschoten, The Netherlands, June.

Lidskog, R. (1993) 'Review of U. Beck, *The Risk Society* [London: Sage, 1992]', *Acta Sociologica* 36(4): 400–3.

Likens, G. and Bormann, F. H. (1974) 'Acid rain: a serious regional environmental problem', *Science* 184, 14 June: 1176–9.

Likens, G. E., Bormann, F. H. and Johnson, N. M. (1972) 'Acid rain', *Environment* 14: 33–40.

Lindblom, C. E. and Cohen, D. H. (1979) *Usable Knowledge: Social Science and Social Problem Solving*. New Haven, CT: Yale University Press.

Lipschutz, R. D. and Conca, K. (eds) (1993) *The State and Social Power in Global Environmental Politics*. New York: Columbia University Press.

Litfin, K. (1993) 'Ecoregimes: playing tug of war with the nation-state', in R. D. Lipschutz and K. Conca (eds) *The State and Social Power in Global Environmental Politics*. New York: Columbia University Press.

Loseke, D. R. (1992) *The Battered Women and Shelters: The Social Construction of Wife Abuse*. Albany: State University of New York Press.

Lovejoy, T. E. (1986) 'Species leave the ark: one by one', in B. G. Norton (ed.) *The Preservation of Species: The Value of Biological Diversity*. Princeton, NJ: Princeton University Press.

Lovely, R. (1990) 'Wisconsin's acid rain battle: science, communication and public policy 1979–1989', *Environmental History Review* 14(3): 20–48.

Lowe, P. D. and Goyder, J. (1983) *Environmental Groups in Politics*. London: Allen & Unwin.

Lowe, P. and Morrison, D. (1984) 'Bad news or good news: environmental politics and the mass media', *The Sociological Review* 32(1): 75–90.

Lundquist, L. J. (1978) 'The comparative study of environmental politics: from garbage to gold?', *International Journal of Environmental Studies* 12: 89–97.

Lutts, R. H. (1990) *The Nature Fakers: Wildlife, Science and Sentiment*. Golden, Col: Fulcrum.

Lynch, B. D. (1993) 'The garden and the sea: U.S. Latino environmental discourses and mainstream environmentalism', *Social Problems* 40(1): 108–24.

Lyotard, J. (1984) *The Postmodern Condition*. Manchester: Manchester University Press.

McAuliffe, S. and McAuliffe, K. (1981) *Life for Sale*. New York: Coward, McCann & Geoghegan.

Macdonald, D. (1991) *The Politics of Pollution: Why Canadians Are Failing Their Environment*. Toronto: McClelland & Stewart.

McIntosh, R. P. (1985) *The Background of Ecology: Concept and Theory*. Cambridge: Cambridge University Press.

Mackie, M. (1987) *Constructing Women and Men: Gender Socialization*. Toronto: Holt, Rinehart & Winston of Canada.

McLaughlin, S. B. (1985) 'Effects of air pollution on forests: a critical review', *Journal of the Air Pollution Control Association* 35: 512–34.

McManus, J. (ed.) (1989) 'Planet of the year', *Time*, 2 January.

McNeely, J. A. (1992) Review of R. Tobin, *The Expendable Future: U.S. Politics and the Protection of Biological Diversity* [Durham, NC: Duke University Press, 1991], *Environment* 34(2): 25–7.

McNeely, J. A., Miller, K. R., Reid, W. V., Mittermeier, R. A. and Werner, T. B. (1990a) *Conserving the World's Biological Diversity*. Gland, Switzerland and Washington, DC: World Conservation Union, World Resources Institute, Conservation International, World Wildlife Fund and World Bank.

McNeely, J. A., Miller, K. R., Reid, W. V., Mittermeier, R. A. and Werner, T. B. (1990b) 'Strategies for conserving biodiversity', *Environment* 32(3): 16–20; 36–40.

Mahon, T. (1985) *Charged Bodies: People, Power and Paradox in Silicon Valley*. New York: New American Library.

Manes, C. (1990) *Green Rage: Radical Environmentalism and the Unmaking of Civilization*. Boston, MA: Little, Brown & Co.

Mann, C. C. and Plummer, M. L. (1992) 'The butterfly problem', *The Atlantic Monthly* 229(1), January: 47–70.

Mannion, A. (1993) Review of W. Reid *et al.*, *Biodiversity Prospecting: Using Genetic Resources for Sustainable Development* [Washington, DC: World Resources Institute, 1993], *Environmental Conservation* 20(3): 286–7.

Martell, L. (1994) *Ecology and Society: An Introduction*. Cambridge: Polity Press.

Maslow, A. K. (1954) *Motivation and Personality*. New York: Harper & Row.

Mauss, A. L. (1975) *Social Problems as Social Movements*. Philadelphia, PA: J. B. Lippincott Co.

Mayer, E. L. (1992) 'Environmental racism', *Audubon*, January/February: 30–2.

Mazur, A. (1985) 'The mass media in environmental controversies', in A. Brannigan and S. Goldenberg (eds) *Social Responses to Technological Change*. Westport, CT: Greenwood Press.

Mazur, A. and Lee, J. (1993) 'Sounding the global alarm: environmental issues in the U.S. national news', *Social Studies of Science* 23: 681–720.

Melucci, A. (1989) *Nomads of the Present*, translated and edited by J. Keane and P. Mier. Philadelphia, PA: Temple University Press.

Merchant, C. (1980) *The Death of Nature: Women, Ecology and the Scientific Revolution: A Feminist Reappraisal of the Scientific Revolution.* New York: Harper & Row.

Merchant, C. (1987) 'The theoretical structure of ecological revolutions', *Environmental Review* 11(4): 265–74.

Merton, R. H. and Nisbet, R. A. (1971) *Contemporary Social Problems*, 3rd edn. New York: Harcourt, Brace World.

Milbrath, L. (1989) *Envisioning a Sustainable Society: Learning Our Way Out.* Albany, NY: State University of New York Press.

Millar, C. I. and Ford, L. D. (1988) 'Managing for nature: from genes to ecosystems', *BioScience* 38(7): 456–7.

Miller, G. and Holstein, J. A. (1993) 'Reconsidering social constructionism', in J. A. Holstein and G. Miller (eds*) Reconsidering Social Constructionism: Debates in Social Problems Theory.* New York: Aldine de Gruyter.

Miller, V. D. (1993) 'Building on our past, planning our future: communities of color and the quest for environmental justice', in R. Hofrichter (ed.) *Toxic Struggles: The Theory and Practice of Environmental Justice.* Philadelphia, PA: New Society Publishers.

Mills, C. W. (1959) *The Sociological Imagination.* New York: Oxford University Press.

Milne, A. (1993) 'The perils of green pessimism', *New Scientist* 138 (#1877), 12 June: 34–7.

Milton, K. (1991) 'Interpreting environmental policy: a social scientific approach', *Journal of Law and Society* 18(1): 4–17.

Modavi, N. (1991) 'Environmentalism, state and economy in the United States', *Research in Social Movements, Conflicts and Change* 13: 261–73.

Molina, M. and Rowland, F. S. (1974) 'Stratospheric sink for chlorofluoromethanes: chlorine atom catalysed destruction of ozone', *Nature* 249, 28 June: 810–12.

Molotch, H. (1970) 'Oil in Santa Barbara and power in America', *Sociological Inquiry* 40: 131–44.

Molotch, H. and Lester, M. (1975) 'Accidental news: the great oil spill as local occurrence and national event', *American Journal of Sociology* 81(2): 235–60.

Morrison, D. E., Hornback, H. E. and Warner, W. H. (1972) 'The environmental movement: some preliminary observations and predictions', in W. R. Burch Jr., N. H. Cheek Jr. and L. Taylor (eds) *Social Behavior, Natural Resources and the Environment.* New York: Harper & Row.

Motavelli, J. (1995) 'Patches of green: environmental programs dot the landscape of TV's "vast wasteland"', *E: The Environmental Magazine* 6(1): 39.

Munt, I. (1994) 'The "other" postmodern tourism: culture, travel and the new middle classes', *Theory, Culture & Society* 11: 101–23.

Murch, A. W. (1971) 'Public concern for environmental pollution', *Public Opinion Quarterly* 35: 100–6.

Nash, R. (1967) *Wilderness and the American Mind.* New Haven, CT: Yale University Press.

Nash, R. (1977) 'The value of wilderness', *Environmental Review* 1: 14–25.

Nash, R. (1989) *The Rights of Nature: A History of Environmental Ethics*. Madison, WI: University of Wisconsin Press.

Nelkin, D. (1987) 'The culture of science journalism', *Society* 24(6): 17–25.

Nelkin, D. (1989) 'Communicating technological risk: the social construction of risk perception', *Annual Review of Public Health* 10: 95–113.

New York Times (1990) 'U.S. sued over drug to increase milk output', *New York Times*, 15 November: A-25.

Norton, B. G. (1986) 'Introduction to Part 1', in B. G. Norton (ed.) *The Preservation of Species: The Value of Biological Diversity*. Princeton, NJ: Princeton University Press.

Noss, R. (1990) 'From endangered species to biodiversity', in K. Kohm (ed.) *Balancing on the Brink of Extinction*. Washington, DC: Island Press.

Novek, J. and Kampen, K. (1992) 'Sustainable or unsustainable development? An analysis of an environmental controversy', *Canadian Journal of Sociology* 17: 249–73.

Oden, S. (1968) 'The acidification of air and precipitation and its consequences in the natural environment', *Ecology Committee Bulletin*, No. 1, Swedish National Research Council.

Oden, S. (1976) 'The acidity problem – an outline of concepts', Proceedings of the First International Symposium on Acid Precipitation and the Forest Ecosystem, USDA General Technical Report NE 23.

Ostmann, R. (1982) *Acid Rain: A Plague Upon the Waters*, Minneapolis, Minn: Dillon Press.

Palmlund, I. (1992) 'Social drama and risk evaluation', in S. Krimsky and D. Golding (eds) *Social Theories of Risk*. Westport, CT: Praeger.

Papadakis, E. (1993) *Politics and the Environment: The Australian Experience*. Sydney: Allen & Unwin.

Park, C. C. (1987) *Acid Rain: Rhetoric and Reality*. London and New York: Methuen.

Park, R. E. (1952) 'Human ecology', in R. E. Park (ed.) *Human Communities: The City and Human Ecology*. New York: The Free Press. (Originally published in 1936 in *American Journal of Sociology* 42: 1–15.)

Parlour, J.W. and Schatzow, S. (1978) 'The mass media and public concern for environmental problems in Canada: 1960–1972', *International Journal of Environmental Studies* 13: 9–17.

Parsons, H. L. (1977) *Marx and Engels on Ecology*. Westport, CT: Greenwood Press.

Pearce, F. (1990) 'Whatever happened to acid rain?', *New Scientist*, 15 September: 57–60.

Pearce, F. (1991) *Green Warriors: The People and Politics Behind the Environmental Revolution*. London: The Bodley Head.

Pearce, F. (1993) 'How Britain hides its acid soil', *New Scientist*, 27 February: 29–33.

Perrow, C. (1984) *Normal Accidents: Living with High Risk Technologies*. New York: Basic Books.

Perry, M. (1994) 'The toilet bowl they call Sydney Harbor', *Toronto Star*, 24 July: B-7.

Pinch, T. J. and Bijker, W. E. (1987) 'The social construction of facts and artifacts: or how the sociology of science and the sociology of technology might benefit one another', in W. E. Bijker, T. P. Hughes and T. J. Pinch (eds) *The Social*

Construction of Technological Systems: New Directions in the Sociology and History of Technology. Cambridge, MA: MIT Press.

Plein, L. C. (1990) 'Biotechnology: issue development and evolution', in D. J. Webber (ed.) *Biotechnology: Assessing Social Impacts and Policy Implications.* Westport, CT: Greenwood Press.

Plein, L. C. (1991) 'Popularizing biotechnology: the influence of issue definition', *Science, Technology & Human Values* 16(4): 474–90.

Quinn, N. (1994) 'Ecologists' credibility going soft', *Toronto Star*, 8 March: A-17.

Rafter, N. (1992) 'Claims-making and socio-cultural context in the first U.S. eugenics campaign', *Social Problems* 35: 17–34.

Redclift, M. (1984) *Development and the Environmental Crisis: Red or Green Alternatives.* New York: Methuen.

Redclift, M. (1986) 'Redefining the environmental "crisis" in the South', in J. Weston (ed.) *Red and Green: The New Politics of the Environment.* London: Pluto Press.

Redclift, M. and Woodgate, G. (1994) 'Sociology and the environment: discordant discourse?', in M. Redclift and T. Benton (eds) *Social Theory and the Global Environment.* London & New York: Routledge.

Regens, J. L. and Rycroft, R. W. (1988) *The Acid Rain Controversy.* Pittsburgh, PA: University of Pittsburgh Press.

Reid, W. V. (1993–4) 'The economic realities of biodiversity', *Issues in Science and Technology* 10(2): 48–55.

Reid, W. V., Laird, S. A., Meyer, C. A., Gamez, R., Sittenfield, A., Jantzen, D. H., Gollin, M. A. and Juma, C. (eds) (1993*)* *Biodiversity Prospecting: Using Genetic Resources for Sustainable Development.* Washington, DC: World Resources Institute.

Renn, O. (1992) 'Concepts of risk: a classification', in S. Krimsky and D. Golding (eds) *Social Theories of Risk.* Westport, CT: Praeger.

Richardson, M., Sherman, J. and Gismondi, M. (1993) *Winning Back the Words: Confronting Experts in an Environmental Public Hearing.* Toronto: Garamond Press.

Rifkin, J. (1983) *Algeny: A New Word – A New World.* New York: The Viking Press.

Rifkin, J. (1992) *Beyond Beef: The Rise and Fall of the Cattle Culture.* New York: Dutton.

Rittenhouse, C. A. (1991) 'The emergence of premenstrual syndrome as a social problem', *Social Problems* 38(3): 412–25.

Rose, C. and Neville, M. (1985) *Tree Dieback Survey: Final Report*, London: Friends of the Earth.

Rosencrantz, A. (1988) 'The acid rain controversy in Europe and North America: a political analysis', in J. E. Carroll (ed.) *International Environmental Diplomacy: The Management and Resolution of Transfrontier Environmental Problems.* Cambridge: Cambridge University Press.

Roth, C. E. (1978) 'Off the merry-go-round on the escalator', in W. B. Stapp (ed.) *From Ought to Action in Environmental Education.* Columbus, OH: ERIE/SMEAC.

Rubin, C. T. (1994) *The Green Crusade: Rethinking the Roots of Environmentalism.* New York: The Free Press.

Russell, A. M. (1988) *The Biotechnology Revolution: An International Perspective.* Sussex: Wheatsheaf Books.

217

Ryan, C. (1991) *Prime Time Activism*. Boston, MA: South End Press.
Rycroft, R. W. (1991) 'Environmentalism and science: politics and the pursuit of knowledge', *Knowledge: Creation, Diffusion, Utilization* 13(2): 150–69.
Salter, L. (with the assistance of E. Levy and W. Leiss) (1988) *Science and Scientists in the Making of Standards*. Dordrecht: Kluwer Academic.
Sandman, P. M., Sachman, D. B., Greenberg, M. and Gochfeld, M. (1987) *Environmental Risk and the Press: An Exploratory Assessment*. New Brunswick, NJ: Transaction Books.
Schlesinger, P. (1978) *Putting 'Reality' Together: BBC News*. London: Constable.
Schmandt, J., Clarkson, J. and Roderick, H. (1988) *Acid Rain and Friendly Neighbors: The Policy Dispute Between Canada and the United States*, revised edn. Durham, NC: Duke University Press.
Schmitt, P. J. (1990) *Back to Nature: The Arcadian Myth in Urban America*. Baltimore, VA: The Johns Hopkins University Press.
Schnaiberg, A. (1980) *The Environment: From Surplus to Scarcity*. New York: Oxford University Press.
Schnaiberg, A. (1993) 'Introduction: inequality once more, with (some) feeling', *Qualitative Sociology* 16(3): 203–6.
Schnaiberg, A. and Gould, K. A. (1994) *Environment and Society*. New York: St Martin's Press.
Schneider, J. W. (1985) 'Social problems theory: the constructionist view', *Annual Review of Sociology* 11: 209–29.
Schneider, J. W. and Kitsuse, J. I. (eds) (1984) *Studies in the Sociology of Social Problems*. Norwood, NJ: Ablex.
Schneider, K. (1989a) 'Gene altered farm drug starts battle in milk states', *New York Times*, 29 April: 1, 8.
Schneider, K. (1989b) 'Vermont resists some progress in dairying', *New York Times*, 27 August: 4.
Schneider, K. (1990a) 'Biotechnology enters political race', *New York Times*, 21 April: A-21.
Schneider, K. (1990b) 'FDA accused of improper ties in review of milk cows', *New York Times*, 12 January: A-21.
Schoenfeld, A. C. (1980) 'Newspapers and the environment today', *Journalism Quarterley* 57: 456–62.
Schoenfeld, A. C., Meier, R. F. and Griffin, R. J. (1979) 'Constructing a social problem: the press and the environment', *Social Problems* 27(1): 38–61.
Schoon, N. (1990) 'Acid rain that earns Britain a black mark', *The Independent*, 26 March: 17.
Schrepfer, S. R. (1983) *The Fight to Save the Redwoods: A History of Environmental Reform 1917–1978*. Madison, WI: The University of Wisconsin Press.
Schumacher, E.F. (1974) *Small is Beautiful*. London: Abacus.
Scotland, R. (1994) 'Marketing model helps rate brand performance', *The Financial Post* (Toronto), 1 December: 23.
Scott, A. (1990) *Ideology and the New Social Movements*. London: Unwin Hyman.
Sears, P. B. (1964) 'Ecology as a subversive subject', *BioScience* 14(7): 11–13.
Selznick, P. (1960) *The Organizational Weapon*. Glencoe, IL: Free Press.
Shapiro, S. (1993) 'Rejoining the battle against noise pollution', *Issues in Science and Technology* 9(3): 73–9.

Sheail, J. (1976) *Nature in Trust: The History of Nature Conservation in Britain*. Glasgow: Blackie.

Shenon. P. (1994) 'A Vietnamese goat is imperiled by fame', *New York Times*, 29 November: A-6.

Shepard, P. (1969) 'Ecology and man: a viewpoint', in P. Shepard and D. McKinly (eds) *The Subversive Science: Essays Toward an Ecology of Man*. Boston, MA: Houghton Mifflin.

Sherlock, R. and Kawar, A. (1990) 'Regulating genetically engineered organisms: the case of the dairy industry', in D. J. Webber (ed.) *Biotechnology: Assessing Social Impacts and Policy Implications*. Westport, CT: Greenwood Press.

Shiva, V. (1990) 'Biodiversity, biotechnology and profit: the need for a People's Plan to protect biological diversity', *The Ecologist* 20(2): 44–7.

Shiva, V. (1993) 'Farmers' rights, biodiversity and international treaties', *Economic and Political Weekly* (Bombay) 28(4), 3 April: 555–60.

Shiva, V. and Holla-Bhar, R. (1993) 'Intellectual piracy and the neem tree', *The Ecologist* 23(6), November/December: 223–7.

Short, J. F. Jr. (1992) 'Refining, explaining and managing risk', in J. F. Short Jr. and L. Clarke (eds) *Organizations, Uncertainties and Risk*. Boulder, CO: Westview Press.

Shove, E. (1994) 'Sustaining developments in environmental sociology', in M. Redclift and T. Benton (eds) *Social Theory and the Global Environment*. London and New York: Routledge.

Simonis, U. (1989) 'Ecological modernization of industrial society: three strategic elements', *International Social Science Journal* 121: 347–61.

Smith, C. (1992) *Media and Apocalypse: News Coverage of the Yellowstone Forest Fires, Exxon Valdez Oil Spill and Loma Prieta Earthquake*. Westport, CT: Greenwood Press.

Smith, R. A. (1872) *Air and Rain: The Beginnings of a Chemical Climatology*. London: Longman, Green.

Smithson, M. (1989) *Ignorance and Uncertainty: Emerging Paradigms*. New York: Springer-Verlag.

Snow, D. A., Rocheford, E. B. Jr., Warden, S. H. and Benford, R. D. (1986) 'Frame alignment processes, micromobilization and movement participation', *American Sociological Review* 51: 464–81.

Solesbury, W. (1976) 'The environmental agenda: an illustration of how situations may become political issues and issues may demand responses from government: or how they may not', *Public Administration* 54: 379–97.

Sorokin, P. A. (1964) [1928] *Contemporary Sociological Theories: Through the First Quarter of the Twentieth Century*. New York: Harper & Row.

Soule, M. and Kohm, K. A. (1989) *Research Priorities for Conservation Biology*. Washington, DC: Island Press.

Spaargaren, G. and Mol, A. P. J. (1992a) 'Sociology, environment and modernity: ecological modernization as a theory of social change', *Society & Natural Resources* 5: 323–44.

Spaargaren, G. and Mol, A. P. J. (1992b) 'Concepts of environment and nature in the context of risk society', Paper presented to the Symposium, 'Current Developments in Environmental Sociology', International Sociological Association, Woudschoten, The Netherlands.

219

Spector, M. and Kitsuse, J. I. (1973) 'Social problems: a reformulation', *Social Problems* 20: 145–59.

Spector, M. and Kitsuse, J. I. (1977) *Constructing Social Problems*. Menlo Park, CA: Cummings.

Spencer, J. W. and Triche, E. (1994) 'Media constructions of risk and safety: differential framings of hazard events', *Sociological Inquiry* 64(2): 199–213.

Spretnak, C. (ed.) (1982) *The Politics of Women's Spirituality: Essays on the Rise of Spiritual Power Within the Women's Movement*. New York: Doubleday/Anchor.

Staggenborg, S. (1993) 'Critical events and the mobilization of the pro-choice movement', *Research in Political Sociology* 6: 319–45.

Stallings, R. (1990) 'Media discourse and the social construction of risk', *Social Problems* 37: 80–95.

Steinmetz, G. (1994) 'Regulation theory, post-Marxism, and the New Social Movements', *Comparative Studies in Society and History* 36(1): 176–212.

Stocking, H. and Holstein, L. W. (1993) 'Constructing and reconstructing scientific ignorance: ignorance claims in science and journalism', *Knowledge: Creation, Diffusion, Utilization* 15:186–210.

Stocking, H. and Leonard, J. P. (1990) 'The greening of the press', *Columbia Journalism Review* 29 (November/December): 37–44.

Susskind, L. E. (1994) *Environmental Diplomacy: Negotiating More Effective Global Agreements*. New York and Oxford: Oxford University Press.

Suzuki, D. (1994a) 'Amazon's "Tarzan" plans epic swim', *Toronto Star*, 4 June: B-1.

Suzuki, D. (1994b) 'Study gives biodiversity a big boost', *Toronto Star*, 5 March: B-6.

Szasz, A. (1994) *Ecopopulism: Toxic Waste and the Movement for Environmental Justice*. Minneapolis: University of Minnesota Press.

Tabara, D. J. (1992) 'Corporate environmental decisions: the case of Barcelona and the Olympic Games Barcelona '92', Paper presented to the Symposium, 'Current Developments in Environmental Sociology', International Sociological Association, Woudschoten, The Netherlands.

Tabara, D. J. and Hannigan, J. A. (1993) 'Urbanization, environmental corporatism and hallmark events: a comparative study', Paper presented to the 31st International Congress of the International Institute of Sociology, The Sorbonne, Paris.

Tangley, L. (1988) 'Research priorities for conservation', *BioScience* 38(7): 444–8.

Tansley, A. G. (1939) 'British ecology during the past quarter century: the plant community and the ecosystem', *Journal of Ecology* 27: 513–30.

Taylor, D. E. (1992) 'Can the environmental movement attract and maintain the support of minorities?', in B. Bryant and P. Mohai (eds) *Race and the Incidence of Environmental Hazards: A Time for Discourse*. Boulder, CO: Westview Press.

Taylor, D. E. (1993) 'Minority environmental activism in Britain: from Brixton to the Lake District', *Qualitative Sociology* 16(3): 263–95.

Taylor, S. (1988) *Patenting Life*. Washington, DC: Congressional Research Service.

Tester, K. (1991) *Animals and Society: The Humanity of Animal Rights*. London and New York: Routledge.

Thompson, M. (1991) 'Plural rationalities: the rudiments of a practical science of the inchoate', in J. Hansen (ed.) *Environmental Concerns: An Interdisciplinary Exercise*. London and New York: Elsevier Applied Science.

Timasheff, N. S. and Theodorson, G. A. (1976) *Sociological Theory: Its Nature and Growth*. New York: Random House.

Tolba, M. K. and El-Kholy, O. A. (1992) *The World Environment 1972–1992: Two Decades of Challenge*. London: Chapman & Hall.

Trop, C. and Roos, L. L. Jr. (1971) 'Public opinion and the environment', in L. L. Roos Jr. (ed.) *The Politics of Ecosuicide*. New York: Holt, Rinehart & Winston.

Tuchman, G. (1978) 'Introduction: the symbolic annihilation of women by the mass media', in G. Tuchman, A. K. Daniels and J. Benet (eds) *Hearth and Home*. New York: Oxford University Press.

Udall, J. R. (1991) 'Launching the natural ark', *Sierra* 76(5), September/October: 80–9.

Ungar, S. (1992) 'The rise and (relative) decline of global warming as a social problem', *The Sociological Quarterly* 33: 483–501.

Ungar, S. (1994) 'Apples and oranges: probing the attitude behaviour relationship for the environment', *Canadian Review of Sociology and Anthropology* 31(3): 288–304.

von Weizsacker, C. (1993) 'Competing notions of biodiversity', in W. Sachs (ed.) *Global Ecology: A New Arena of Political Conflict*. London: Zed Books.

Walker, J. L. (1977) 'Setting the agenda of the U.S. Senate: a theory of problem selection', *British Journal of Political Science* 7: 423–45.

Walker, J. L. (1981) 'The diffusion of knowledge, policy communities and agenda setting: the relationship of knowledge and power', in J. E. Tropman, M. J. Dluhy and R. M. Lind (eds) *New Strategic Perspectives on Social Policy*. New York: Pergamon Press.

Ward, B. (1966) *Spaceship Earth*. New York: Columbia University Press.

Weitzer, R. (1991) 'Prostitutes' rights in the United States: the failure of a movement', *The Sociological Quarterly* 32(1): 23–41.

Welsh, I. (1992) 'Education for what? Environment, ecology and sociology', Paper presented to the Symposium, 'Current Developments in Environmental Sociology', International Sociological Association, Woudschoten, The Netherlands.

Westell, D. (1994) 'Urban treatment of sewage gets bad marks: study shows 17 cities fail to treat all waste', *The Globe & Mail*, 15 June: A-4.

Wharton, C. R. Jr. (1966) 'Modernizing subsistence agriculture', in M. Wiener (ed.) *Modernization: The Dynamics of Growth*. New York: Basic Books.

Wiener, C. L. (1981) *The Politics of Alcoholism: Building an Arena around a Social Problem*. New Brunswick, NJ: Transaction.

Wilcher, M. E. (1989) *The Politics of Acid Rain: Policy in Great Britain and the United States*. Aldershot: Avebury.

Wilkins, L. and Patterson, P. (1990) 'Risky business: covering slow-onset hazards as rapidly developing news', *Political Communication and Persuasion* 7(1): 11–23.

Wilson, A. (1992) *The Culture of Nature: North American Landscape from Disney to the Exxon Valdez*, Cambridge, MA: Blackwell.

Wilson, E.O. (1986) 'Editor's foreword', in E.O. Wilson (ed.) *Biodiversity*. Washington, DC: National Academy Press.

Wilson, E. O. (ed.) (1988) *Biodiversity*. Washington, DC: National Academy Press.

Wilson, E. O. (1994) *Naturalist*. Washington, DC and Cavelo, CA: Island Press.

Witt, W. (1974) 'The environmental reporter on U.S. daily newspapers', *Journalism Quarterly* 51: 697–704.

Woolgar, S. and Pawluch, D. (1985) 'Ontological gerrymandering', *Social Problems* 32(3): 214–37.

Worster, D. (1977) *Nature's Economy: A History of Ecological Ideas*. Cambridge: Cambridge University Press.

Wright, B. and Weiss, J. P. (1980) *Social Problems*. Boston: Little, Brown & Co.

Wynne, B. (1992) 'Risk and social learning: reification to engagement', in S. Krimsky and D. Golding (eds) *Social Theories of Risk*. Westport, CT: Praeger.

Wynne, B. and Mayer, S. (1993) 'How science fails the environment', *New Scientist* 138 (#1876), 5 June: 33–5.

Yanarella, E. J. (1985) 'The foundations of policy immobilism over acid rain control', in E. J. Yanarella and R. H. Ihara (eds) *The Acid Rain Debate: Scientific, Economic and Political Dimensions*. Boulder, CO: Westview Press.

Yearley, S. (1992) *The Green Case: A Sociology of Environmental Issues, Arguments and Politics*. London: Routledge.

Zehr, S. C. (1994) 'The centrality of scientists and the translation of interests in the U.S. acid rain controversy', *Canadian Review of Sociology & Anthropology* 31 (3): 325–53.

Zukin, S. (1988) 'Postmodern urban landscapes: mapping culture and power', in S. Lash and J. Friedman (eds) *Modernity and Identity*. Oxford: Blackwell.

NAME INDEX

SUBJECT INDEX